MODELING OF DIGITAL COMMUNICATION SYSTEMS USING SIMULINK®

MODELING OF DIGITAL COMMUNICATION SYSTEMS USING SIMULINK®

ARTHUR A. GIORDANO & ALLEN H. LEVESQUE

WILEY

Copyright © 2015 by John Wiley & Sons, Inc. All rights reserved

Published by John Wiley & Sons, Inc., Hoboken, New Jersey
Published simultaneously in Canada

MATLAB and Simulink are registered trademarks of The MathWorks, Inc. See www.mathworks.com/ trademarks for a list of additional trademarks. **The MathWorks Publisher Logo identifies books that contain MATLAB® content. Used with permission. The MathWorks does not warrant the accuracy of the text or exercises in this book or in the software downloadable from** http://www.wiley.com/WileyCDA/WileyTitle/ productCd-047064477X.html **and** http://www.mathworks.com/matlabcentral/fileexchange/?term=authorid% 3A80973. **The book's or downloadable software's use or discussion of MATLAB® software or related products does not constitute endorsement or sponsorship by The MathWorks of a particular use of the MATLAB® software or related products.**

For MATLAB® and Simulink® product information, or information on other related products, please contact:

The MathWorks, Inc.
3 Apple Hill Drive
Natick, MA 01760-2098 USA
Tel 508-647-7000
Fax: 508-647-7001
E-mail: info@mathworks.com
Web: www.mathworks.com
How to buy: www.mathworks.com/store

No part of this publication may be reproduced, stored in a retrieval system, or transmitted in any form or by any means, electronic, mechanical, photocopying, recording, scanning, or otherwise, except as permitted under Section 107 or 108 of the 1976 United States Copyright Act, without either the prior written permission of the Publisher, or authorization through payment of the appropriate per-copy fee to the Copyright Clearance Center, Inc., 222 Rosewood Drive, Danvers, MA 01923, (978) 750-8400, fax (978) 750-4470, or on the web at www.copyright.com. Requests to the Publisher for permission should be addressed to the Permissions Department, John Wiley & Sons, Inc., 111 River Street, Hoboken, NJ 07030, (201) 748-6011, fax (201) 748-6008, or online at http://www.wiley.com/go/permission.

Limit of Liability/Disclaimer of Warranty: While the publisher and authors have used their best efforts in preparing this book, they make no representations or warranties with respect to the accuracy or completeness of the contents of this book and specifically disclaim any implied warranties of merchantability or fitness for a particular purpose. No warranty may be created or extended by sales representatives or written sales materials. The advice and strategies contained herein may not be suitable for your situation. You should consult with a professional where appropriate. Neither the publisher nor authors shall be liable for any loss of profit or any other commercial damages, including but not limited to special, incidental, consequential, or other damages.

For general information on our other products and services or for technical support, please contact our Customer Care Department within the United States at (800) 762-2974, outside the United States at (317) 572-3993 or fax (317) 572-4002.

Wiley also publishes its books in a variety of electronic formats. Some content that appears in print may not be available in electronic formats. For more information about Wiley products, visit our web site at www.wiley.com.

MATLAB® is a trademark of The MathWorks, Inc. and is used with permission. The MathWorks does not warrant the accuracy of the text or exercises in this book. This book's use or discussion of MATLAB® software or related products does not constitute endorsement or sponsorship by The MathWorks of a particular pedagogical approach or particular use of the MATLAB® software.

Library of Congress Cataloging-in-Publication Data:

Giordano, Arthur A. (Arthur Anthony), 1941-
  Modeling of digital communications systems using Simulink / Arthur A. Giordano & Allen H. Levesque.
    pages cm
  Includes bibliographical references and index.
  ISBN 978-1-118-40005-0 (cloth)
  1. Digital communications–Computer simulation. 2. SIMULINK. I. Levesque, Allen H. II. Title.
  TK5102.83.G56 2015
  621.3820285′53–dc23
                                                                                            2014042283

Cover Image courtesy of iStockphoto © Greyfebruary

Typeset in 11/13pt Times by Laserwords Private Limited Chennai, India

10 9 8 7 6 5 4 3 2 1

1 2015

To our families, especially Diane and Barbara

CONTENTS

Preface	xiii
Acknowledgments	xix
About the Companion website	xxi
Abbreviations and Acronyms	xxiii

1 **Getting Started with Simulink**   1

    1.1   Introduction, 1
    1.2   Starting a Matlab Session, 2
    1.3   Simulink Block Libraries, 3
    1.4   Building a New Simulink Model, 6
          1.4.1   Inserting Signal Source and Scope, 6
          1.4.2   Setting the Source Block Parameters, 8
          1.4.3   Setting Scope Parameters, 9
    1.5   Executing the Simulink Model, 11
    1.6   Reconfiguring the Signal Block, 14
    1.7   Sample-Based Signals, 16
    1.8   Sending Data to Workspace, 18
    1.9   Using Model Explorer, 19
    1.10 Adding Labels to Figures, 21
    1.11 Selecting Model Configuration Parameters, 22
    1.12 Summary Discussion, 24
          Problems, 25

2 Sinusoidal Simulink Model — 27

- 2.1 A First Simulink Model, 27
- 2.2 Simulink Model of Sine Wave, 27
- 2.3 Spectrum of a Sine Wave, 32
- 2.4 Summary Discussion, 40
- Problems, 41

3 Digital Communications BER Performance in AWGN (BPSK and QPSK) — 43

- 3.1 BPSK and QPSK Error Rate Performance in AWGN, 43
- 3.2 Construction of a Simulink Model in Simple Steps, 44
- 3.3 Comparison of Simulated and Theoretical BER, 56
- 3.4 Alternate Simulink Model for BPSK, 58
- 3.5 Frame-Based Simulink Model, 62
- 3.6 QPSK Symbol Error Rate Performance, 64
- 3.7 BPSK Fixed Point Performance, 68
- 3.8 Summary Discussion, 73
- Appendix 3.A Theoretical BER Performance of BPSK in AWGN, 73
- Problems, 75

4 Digital Communications BER Performance in AWGN (MPSK & QAM) — 79

- 4.1 MPSK and QAM Error Rate Performance in AWGN, 79
- 4.2 MPSK Simulink Model, 79
- 4.3 BER for Other Alphabet Sizes, 83
- 4.4 Fixed Point BER for MPSK, 83
- 4.5 QAM Simulink Model, 85
- 4.6 QAM BER for Other Alphabet Sizes Using Average Power, 90
- 4.7 QAM BER Using Peak Power, 90
- 4.8 Power Amplifier Constraint Using Peak Power Selection with QAM, 91
- 4.9 Summary Discussion, 99
- Problems, 100

5 Digital Communications BER Performance in AWGN (FSK and MSK) — 101

- 5.1 FSK and MSK Error Rate Performance in AWGN, 101
- 5.2 BFSK Simulink Model, 101

5.3 MFSK Simulink Model, 107
5.4 MSK Simulink Model, 108
5.5 MSK Power Spectrum, 113
5.6 Summary Discussion, 116
Problems, 117

6 Digital Communications BER Performance in AWGN (BPSK in Fading) 119

6.1 BPSK in Rayleigh and Rician Fading, 119
6.2 BPSK BER Performance in Rayleigh Fading, 119
6.3 BPSK BER Performance in Rician Fading, 124
6.4 BPSK BER Performance in Rician Fading with Multipath, 127
6.5 Summary Discussion, 137
Appendix 6.A Theoretical BER Performance of BPSK in Rayleigh Fading, 137
Appendix 6.B Theoretical BER Performance of BPSK in Rician Fading, 138
Problems, 139

7 Digital Communications BER Performance in AWGN (FSK in Fading) 141

7.1 FSK in Rayleigh and Rician Fading, 141
7.2 BFSK BER Performance in Rayleigh Fading, 141
7.3 MFSK BER Performance in Rayleigh Fading, 142
7.4 BFSK BER Performance in Rician Fading, 147
7.5 BFSK BER Performance in Rician Fading with Multipath, 148
7.6 Summary Discussion, 150
Appendix 7.A Theoretical BER Performance of FSK in Rayleigh and Rician Fading, 152
Rayleigh Fading, 152
Rician Fading, 153
Problems, 154

8 Digital Communications BER Performance (STBC) 157

8.1 Digital Modulations in Rayleigh Fading with STBC, 157
8.2 BPSK BER in Rayleigh Fading with STBC, 157
8.3 QAM BER in Rayleigh Fading with STBC, 163
8.4 Summary Discussion, 163
Appendix 8.A Space–Time Block Coding for BPSK, 165

Appendix 8.B Space–Time Block Coding for 16-QAM, 167
Problems, 169

9 Digital Communications BER Performance in AWGN (Block Coding)   **171**

9.1 Digital Communications with Block Coding in AWGN, 171
9.2 BER Performance of BPSK in AWGN with a Binary BCH Block Code, 171
9.3 BER Performance of BPSK in AWGN with a Hamming Code, 175
9.4 BER Performance of BPSK in AWGN with a Golay(24,12) Block Code, 179
9.5 BER Performance of FSK in AWGN with Reed-Solomon Code, 181
9.6 BER Performance of QAM in AWGN with Reed-Solomon Coding, 186
9.7 Summary Discussion, 190
Problems, 192

10 Digital Communications BER Performance in AWGN (Block Coding and Fading)   **193**

10.1 Digital Communications with Block Coding in Fading, 193
10.2 BER Performance of BPSK in Rayleigh Fading with Interleaving and a BCH Block Code, 194
10.3 BER Performance of BFSK in Rayleigh Fading with Interleaving and a Golay(24,12) Block Code, 195
10.4 BER Performance of 32-FSK in Rayleigh Fading with Interleaving and a Reed-Solomon(31,15) Block Code, 201
10.5 BER Performance of 16-QAM in Rayleigh Fading with Interleaving and a Reed-Solomon(15,7) Block Code, 204
10.6 BER Performance of 16-QAM in Rayleigh and Rician Fading with Interleaving and a Reed-Solomon(15,7) Block Code, 208
10.7 BER Performance of BPSK in Rayleigh Fading with Interleaving and a BCH Block Code and Alamouti STBC, 210
10.8 BER Performance of BFSK in Rayleigh Fading with Interleaving and a Golay(24,12) Block Code and Alamouti STBC, 215
10.9 BER Performance of 32-FSK in Rayleigh Fading with Interleaving and a Reed-Solomon(31,15) Block Code and Alamouti STBC, 218

10.10 BER Performance of 16-QAM in Rayleigh Fading with Interleaving and a Reed-Solomon (15,7) Block Code and Alamouti STBC, 219
10.11 Summary Discussion, 223
Problems, 224

11 Digital Communications BER Performance in AWGN and Fading (Convolutional Coding) — 225

11.1 Digital Communications with Convolutional Coding in AWGN and Fading, 225
11.2 BER Performance of Convolutional Coding and BPSK in AWGN, 226
    11.2.1 Hard-Decision Decoding, 226
    11.2.2 Soft-Decision Decoding, 229
11.3 BER Performance of Convolutional Coding and BPSK in AWGN and Rayleigh Fading with Interleaving (Soft- and Hard-Decision Decoding), 233
11.4 BER Performance of Convolutional Coding and BPSK and Alamouti STBC in Rayleigh Fading with Interleaving, 239
11.5 Summary Discussion, 243
Problems, 244

12 Adaptive Equalization in Digital Communications — 247

12.1 Adaptive Equalization, 247
12.2 BER Performance of BPSK in Dispersive Multipath Channel Using an LMS Linear Equalizer, 248
12.3 BER Performance of BPSK in Dispersive Multipath Channel Using an LMS Linear Equalizer From the Simulink Library, 257
12.4 BER Performance of QPSK in a channel with ISI Using an LMS Linear Equalizer, 258
12.5 BER Performance of BPSK in Dispersive Multipath Channel Using a Decision Feedback Equalizer, 268
12.6 BER Performance of BPSK in Rayleigh Fading Multipath Channel Using an RLS Equalizer, 273
    12.6.1 RLS Equalizer Description, 273
    12.6.2 RLS Equalization in Rayleigh Fading with No Multipath, 275
    12.6.3 RLS Equalization in Rayleigh Fading with Multipath, 279

12.7 Summary Discussion, 280
Problems, 283

13 Simulink Examples — 285

13.1 Linear Predictive Coding (LPC) for Speech Compression, 286
    13.1.1 Speech Vocal Tract Model, 289
    13.1.2 Prediction Coefficients Computation, 289
    13.1.3 Speech Analysis and Synthesis, 289

13.2 RLS Interference Cancellation, 291
    13.2.1 Sinusoidal Interference, 291
    13.2.2 Low Pass Filtered Gaussian Noise, 296

13.3 Spread Spectrum, 298
    13.3.1 Spread Spectrum Simulink Model without In-Band Interference, 298
    13.3.2 Spread Spectrum Simulink Model with In-Band Interference, 303
    13.3.3 Spread Spectrum Simulink Model with In-Band Interference and Excision, 309

13.4 Antenna Nulling, 313

13.5 Kalman Filtering, 320
    13.5.1 Scalar Kalman Filter, 322
    13.5.2 Kalman Equalizer, 328
    13.5.3 Radar Tracking Using Extended Kalman Filter (EKF), 339

13.6 Orthogonal Frequency Division Multiplexing, 343

13.7 Turbo Coding with BPSK, 355

Appendix A   Principal Simulink Blocks Used In Chapters 1–13 — 363

Appendix B   Further Reading — 369

Index — 371

PREFACE

This book is designed to introduce the communications systems engineer to the use of The MathWorks® Simulink®[1] for modeling and evaluating the performance of digital communications systems. Simulink is a block-oriented modeling tool that utilizes well-tested MATLAB® code to enable rapid development of simulations for communication systems modeling. This block-oriented approach obviates the need for writing new software routines. The Simulink library provides an extensive array of MathWorks-verified blocks available for assembling any specific model. Upon gaining facility with the use of Simulink, the user will have a robust engineering tool for estimating communication systems performance in instances where analytic results are unavailable, such as nonlinear systems or time-varying channels.

---

[1] The MathWorks™ is the leading developer of mathematical computing software for engineers and scientists. Founded in 1984, with headquarters in Natick, Massachusetts, USA. Simulink® is a block diagram environment for multidomain simulation and Model-Based Design. It supports simulation, automatic code generation, and continuous test and verification of embedded systems.

SCOPE

This book introduces the reader to Simulink, an extension of the widely-used MATLAB[2] modeling tool, and the use of Simulink in modeling and simulating digital communication systems, including wireless communications systems. In contrast with other books that treat MATLAB in depth but treat Simulink only at an introductory level, this book enables the communication systems engineer to learn and use the extensive capabilities of Simulink to model a wide selection of digital communications systems and evaluate their performance for many important channel conditions.

The reader is expected to have an understanding of MATLAB and its environment. It is also expected that the reader has a basic knowledge of digital communications including modulation, coding, and channel models, digital signal processing (DSP) such as digital filtering and Fourier transforms and statistical communications. The book is not intended to be a substitute for a course in digital communications but can be a valuable accompaniment to such a course. The presentation in this text is designed to allow the user to gain familiarity with basic Simulink tools, enabling rapid construction of useful communications systems models rather than providing comprehensive Simulink training currently available from The MathWorks.

Another feature of Simulink, not treated in this book, is Simulink's capability to develop a model, produce C/C++ code and migrate the model for incorporation in an field-programmable gate array (FPGA) or DSP devices. The MathWorks Corporation provides training for this capability.

ORGANIZATION OF THE BOOK

The book is organized in two parts. The first 12 chapters address Simulink models of digital communication systems using various modulation, coding, channel type, and receiver processing techniques. These chapters include theoretical results for known conditions, when available, and simulated results in other cases. Problem sets at the end of each chapter accompany topics to be emphasized. Chapter 13 provides an extensive collection of examples designed to acquaint the reader with applications that reveal the power of Simulink modeling. Appendix A summarizes principal Simulink blocks used in chapters 1–13. Appendix B provides a list of references for further reading on MATLAB and Simulink.

---

[2]MATLAB is a high-level language and interactive environment for numerical computation, visualization, and programming. Computations are most generally performed using vector and/or matrix representations.

CHAPTER 1

In this chapter, the fundamentals of developing a Simulink model and its relationship to MATLAB are described. Screens encountered in a typical MATLAB session are presented.[3] The Simulink library blocks are identified with a focus on the Communications System Toolbox and the DSP System Toolbox.

CHAPTER 2

This chapter is intended to introduce the user to the first and simplest Simulink model associated with a sinusoidal waveform. Block parameters are identified, the simulation is performed and outputs are sent to the Workspace. Blocks that display data and scopes showing waveforms are employed. A Fast Fourier Transform (FFT) block is used to compute the spectrum and compare with a sinusoid's known spectrum. Two blocks are used to determine the spectrum including a spectrum analyzer that has multiple spectrum settings. Commonly used math blocks are also incorporated into the Simulink model.

CHAPTER 3

This chapter introduces several topics in Simulink based on binary phase shift keying (BPSK) and quadrature phase shift keying (QPSK) modulations. Communications blocks for BPSK and QPSK are utilized along with communications channel blocks identified as additive white Gaussian noise (AWGN) and Gaussian noise. BPSK and QPSK bit error rate (BER) performance is simulated and compared with the corresponding theoretical BER results. Use of the bertool is shown to be a convenient technique to obtain BER performance over a range of bit energy to noise spectral density values. Sample-based and frame-based computations are presented where it is seen that frame-based computations are vector based and provide faster computation. The chapter concludes with Simulink computations employing fixed-point arithmetic.

CHAPTER 4

This chapter continues the development of Simulink models for coherent modulations including multi-phase PSK (MPSK) and quadrature amplitude

---

[3]Matlab screens are Reprinted with permission from The MathWorks, Inc.

modulation (QAM). Simulink modeling of BER performance for MPSK using floating- and fixed-point arithmetic is obtained for various alphabet sizes. Simulink modeling of QAM BER is performed for both average and peak power conditions, again with various alphabet sizes. Using an example of QAM signaling in conjunction with a nonlinear power amplifier highlights the power of Simulink to model a communication system and determine its performance in a case where theoretical results are not available.

CHAPTER 5

This chapter continues the development of Simulink models to determine simulated BER results focusing on binary frequency shift keying (BFSK), M-ary frequency shift keying (MFSK), minimum shift keying (MSK), and Gaussian minimum shift keying (GMSK). Comparison of simulated and theoretical performance confirms the facility to reliably estimate performance. The ability of the spectrum analyzer to exhibit a wide selection of spectral estimation techniques and parameters is demonstrated. Spectral efficiencies of BFSK, 4-FSK, MSK, and GMSK are obtained.

CHAPTER 6

Prior chapters have introduced Simulink models to acquaint the reader and obtain confidence in the results from Simulink model development. This chapter presents several topics in Simulink based on BPSK modulation in fading channels. Specifically both Rayleigh and Rician fading channels are incorporated in the Simulink models and used to simulate BPSK BER performance under a variety of channel conditions. Comparisons of theoretical with simulated results are performed. Simulink models that introduce multipath allow BPSK BER performance to be readily estimated and are examples where BER performance is not easily obtained analytically.

CHAPTER 7

Chapter 7 continues the investigation of BER performance using Simulink models incorporating Rayleigh and Rician fading channels with FSK modulation and noncoherent detection. BER performance for Simulink models implementing MFSK in Rayleigh fading is determined for a selection of alphabet sizes. This chapter concludes with a Simulink model that investigates the BER performance of coherently detected FSK in a multipath channel with Rician fading.

CHAPTER 8

Use of diversity is a well-known technique for mitigating the loss in performance due to fading. Space time block coding (STBC) using multiple transmit and/or receive antennas can substantially improve BER performance in these instances. The Simulink models presented here incorporate STBC for compensating for BER degradation due to channel fading. Using STBC, BER performance for BPSK and QAM modulations in Rayleigh fading is determined where it is seen that Simulink obviates the need for theoretical results.

CHAPTER 9

Block error control coding is an important technique to enhance communication system performance. Simulink models for BPSK in AWGN are used to develop BER performance for operation with common block codes including Hamming, Golay, and Bose–Chadhuri–Hocquenghem (BCH). Simulink models are also developed for FSK and QAM with Reed-Solomon codes. A Simulink example is provided to demonstrate the degradation due to multipath with and without coding.

CHAPTER 10

This chapter presents topics in Simulink based on block error control coding in a fading channel. Simulink models employing Rayleigh fading are shown for BCH coding with BPSK, Golay coding with BFSK, Reed-Solomon coding with 32-FSK, and Reed-Solomon coding with 16-QAM. In each of these cases, an interleaver is introduced. A Simulink model is also developed for Rician fading and interleaving using Reed-Solomon coding with 16-QAM. A concluding section presents Simulink models for selected block codes and modulations with STBC and interleaving in Rayleigh fading.

CHAPTER 11

Convolutional coding is another technique utilized to enhance communications system performance. This chapter presents topics in Simulink incorporating convolutional error control coding in an AWGN and a fading channel. Simulink models are developed for computing BER performance of convolutional coding and BPSK in AWGN and Rayleigh fading using both hard- and soft-decision decoding. This chapter concludes with a Simulink model for determining BER performance of convolutional coding and BPSK with STBC and interleaving in Rayleigh fading.

CHAPTER 12

Adaptive equalization has been used extensively to compensate for the degradations from time- dispersive multipath channels. Simulink models, incorporating adaptive equalization for multipath mitigation, are developed for linear least mean square (LMS) equalizers with BPSK and QPSK, decision feedback equalizers (DEFs) with BPSK and recursive least squares (RLS) equalizers with BPSK. A final Simulink model in this chapter was developed using RLS equalization with BPSK in Rayleigh fading. Simulation is required in each of these cases and Simulink is the tool that achieves the desired results.

CHAPTER 13

This chapter provides several Simulink examples for a variety of individual applications. In particular Simulink models are developed for the following situations: Linear Predictive Coding (LPC) for speech compression, RLS interference cancellation, spread spectrum, antenna nulling of a single interferer, Kalman filtering, Orthogonal Frequency Division Multiplexing (OFDM), and Turbo Coding with BPSK. The choice of topics is meant to illustrate the multiplicity of applications that can be investigated using Simulink modeling.

ABOUT THE SOFTWARE

Simulink models for all of the cases presented in this book are available on a companion website, www.wiley.com/go/simulink. All of the models have been successfully executed in MATLAB 2014a. The website also provides Simulink models for the problem sets, and for instructors, answers to the problems can be obtained through the website.

ACKNOWLEDGMENTS

This book was made possible by the generous support from The MathWorks™, who provided multiple MATLAB®/Simulink® releases during the book's preparation. The MathWorks technical staff provided valuable help in identifying and correcting Simulink model issues and for this we are extremely grateful. MATLAB Central is also a significant source of user-developed Simulink models that often enable the developer to obtain a satisfactory resolution of a complex problem. We want to thank Dr. Michael Mulligan at The MathWorks™ for agreeing to support this undertaking. We want to especially thank The MathWorks™ for offering to include our work in The MathWorks Book Program and for granting permissions to reprint a number of copyrighted figures throughout the book.

ABOUT THE COMPANION WEBSITE

This book is accompanied by a companion website:
http://www.wiley.com/go/simulink

The website includes:
- Solutions Manual available to Instructors.

ABBREVIATIONS AND ACRONYMS

ACF	autocorrelation function
AGC	automatic gain control
AM	amplitude modulation
ASIC	application-specific integrated circuit
AWGN	additive white Gaussian noise
BCH	Bose–Chaudhuri–Hocquenghem
BER	bit error rate
BFSK	binary frequency-shift keying
BPSK	binary phase-shift keying
BSC	binary symmetric channel
CPFSK	continuous phase frequency-shift keying
dB	decibel
DFE	decision feedback equalizer
DSB	double-sideband
DSP	digital signal processing
EKF	extended Kalman filter
FIR	finite impulse response
FFT	fast Fourier transform
FSK	frequency-shift keying
FPGA	field-programmable gate array
FM	frequency modulation
GMSK	Gaussian minimum-shift keying

GSM	Global System for Mobile Communications
IFFT	inverse fast Fourier transform
ISI	intersymbol interference
LMS	least mean square
LPC	linear predictive coding
LSB	least significant bit
LTE	Long-Term Evolution
MIMO	multiple-input multiple-output
MFSK	M-ary frequency-shift keying (also multiple FSK)
MPSK	Multiphase PSK (also M-ary PSK)
MQPSK	M-ary quadrature phase-shift keying
MSB	most significant bit
MSE	mean square error
MSK	minimum-shift keying
OFDM	orthogonal frequency division multiplexing
OQPSK	offset quaternary (also quadrature) phase-shift keying
OSTBC	orthogonal space-time block coding
PA	power amplifier
PCCC	parallel concatenated convolutional code
PM	phase modulation
PSK	phase-shift keying
QAM	quadrature amplitude modulation
QPSK	quaternary phase-shift keying (also quadrature PSK)
RBW	resolution bandwidth
RC	raised-cosine
RLS	recursive least squares
RS	Reed–Solomon
SER	symbol error rate
SISO	soft input soft output
SNR	signal-to-noise ratio
SSPA	solid-state power amplifier
STBC	space-time block coding
TWT	traveling-wave tube
VA	Viterbi algorithm
WIMAX	Worldwide Interoperability for Microwave Access

1

GETTING STARTED WITH SIMULINK

1.1 INTRODUCTION

This chapter describes the basic steps to be followed in building a Simulink model. The model presented in this chapter is a simple one: generation of a sinusoidal signal. This presentation uses MATLAB 2014a[1], incorporating Simulink and blocks extracted from the Simulink Library Browser. Subsequent chapters will utilize blocks from the Simulink Library Browser including the Communications System Toolbox and the DSP System Toolbox. The Communications System Toolbox provides a collection of MATLAB functions and simulation blocks that can be utilized for a wide range of digital communications systems simulation models. While MATLAB and Simulink are available for a variety of operating systems, all of the descriptions and examples presented in this book are implemented on Windows-based computers.

A comprehensive presentation of MATLAB® and Simulink® by The MathWorks™ is available at http://www.mathworks.com/help/. Product descriptions from this documentation are provided as follows:

---

[1]Every model presented in this book executes successfully in MATLAB 2014a. In some instances, a notice is generated indicating that the model was developed in an earlier release.

Modeling of Digital Communication Systems Using SIMULINK®, First Edition.
Arthur A. Giordano and Allen H. Levesque.
© 2015 John Wiley & Sons, Inc. Published 2015 by John Wiley & Sons, Inc.
Companion Website: www.wiley.com/go/simulink

MATLAB®

MATLAB® is a high-level language and interactive environment for numerical computation, visualization, and programming. Using MATLAB, you can analyze data, develop algorithms, and create models and applications. The language, tools, and built-in math functions enable you to explore multiple approaches and reach a solution faster than with spreadsheets or traditional programming languages, such as C/C++ or Java®. MATLAB is an abbreviation for "matrix laboratory." While other programming languages mostly work with numbers one at a time, MATLAB is designed to operate primarily on whole matrices and arrays.

Simulink®

Simulink® is a block diagram environment for multidomain simulation and Model-Based Design. It supports system-level design, simulation, automatic code generation, and continuous test and verification of embedded systems. Simulink provides a graphical editor, customizable block libraries, and solvers for modeling and simulating dynamic systems. It is integrated with MATLAB, enabling you to incorporate MATLAB algorithms into models and export simulation results to MATLAB for further analysis.

Upon gaining familiarity with Simulink, the user will discover that multiple paths can be followed in developing a Simulink model. The choice of the path is left to the user but each path will lead to the same solution.

The topics covered in this chapter are:

- Starting a MATLAB session
- Viewing Simulink block libraries
- Building a new Simulink model
  - Setting simulation parameters
  - Setting and using Scopes
- Executing the model
- Sending data to Workspace
- Using the Model Explorer
- Selecting Model Configuration Parameters

1.2 STARTING A MATLAB SESSION

To start a MATLAB session on a Windows machine, simply double-click on the MATLAB icon; this will open the MATLAB desktop, shown in Figure 1.1.

The default desktop view shown in this text includes four panels. (Other desktop views may be selected under the Home menu; see the Layout tabs on the toolbar.) The left-hand panel displays the Current Folder, which can contain MATLAB or Simulink models in addition to user-developed figures and user files resident in the folder. The center panel is the Command Window, where the user inserts MATLAB commands, assigns values to model parameters, and performs calculations using MATLAB mathematical functions. The upper right-hand panel is the Workspace, where variables defined in the Command Window are displayed, and the lower right-hand panel is the Command History, where the user can view or rerun commands entered at the command line. In Figure 1.1 the bar above the Command Window is labeled here as
▶ C: ▶ Users ▶ Default User

The path can be changed to a different folder. At the end of this bar the symbol ▼ also allows the path to be changed to a previous selection.

1.3 SIMULINK BLOCK LIBRARIES

Building a Simulink model consists of selecting individual blocks contained in libraries and joining them in a block diagram of the system to be simulated.

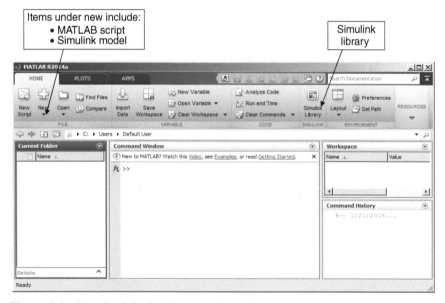

Figure 1.1 MATLAB Default Desktop View. Reprinted with permissions from The MathWorks™, Inc[2].

[2]This Simulink parameter window and similar figures appearing throughout the book are all reprinted with permissions from The MathWorks™, Inc.

Figure 1.2 Simulink Block Library.

To view available blocks, select Simulink Library on the MATLAB toolbar. This opens the window shown in Figure 1.2.

The Simulink Library Browser shows a listing of available Simulink blocks. The focus in this book is on modeling digital communication systems, and the blocks you will find most useful are contained in the basic Simulink block library as well as in the Communications System Toolbox and the DSP System Toolbox.

Simulink library blocks used throughout this book are listed here.

Simulink

- **Commonly used blocks**
    - Constant
    - Delay
    - In1 & Out1
    - Scope
    - Math operations
        - Abs
        - Sum
        - Product
        - Complex to Real-Imag
        - Math Function
    - Model Wide Utilities

- Model Info
  - Signal Routing
    - From
    - Goto
    - Mux
  - Sinks
    - Display
    - Scope
    - To Workspace
  - Sources
    - Constant
    - From Workspace
    - Random number
    - Sine Wave
  - User-Defined Functions
    - Matlab Function
- **Communications System Toolbox**
  - Channels
    - AWGN
    - Multipath Rayleigh Fading
  - Comm Sources
    - Noise Generators
      - Gaussian Noise Generator
    - Random Data Sources
      - Bernoulli Binary Generator
      - Random Integer Generator
  - Error Detection & Correction
    - Block
    - Convolutional
  - Modulation
    - Digital Baseband Modulation
      - AM (QAM)
      - PM (BPSK,QPSK,M-PSK)
      - FM(M-FSK)
- **DSP System Toolbox**
  - Filtering

- Adaptive Filters
  - Block LMS Filter
  - Kalman Filter
  - RLS Filter
- Signal Management
  - Buffers
  - Sinks
    - Spectrum Analyzer
    - Time Scope
    - Vector Scope
  - Sources (DSP Sine Wave)
  - Statistics
    - Mean
    - Variance
    - Autocorrelation
    - Correlation
  - Transforms
    - FFT
    - IFFT

Appendix A lists principal Simulink blocks used in Chapters 1–13.

1.4 BUILDING A NEW SIMULINK MODEL

To begin building a new Simulink model, on the MATLAB toolbar, under the **HOME** tab, pull down **New** and select **Simulink Model**. This will open a blank Simulink model window, shown in Figure 1.3. Note that on the title bar at the top of the window, this model is labeled untitled. In the model window, the user may select the duration of the model execution, shown here to be set at 10.0 s. This will fix the duration of each of the simulations to be demonstrated in this chapter.

To rename the model, on the toolbar, select **File or File:save as** and enter the model name **First_Simulink_Model**, then **save**. The renamed model is shown in Figure 1.4.

1.4.1 Inserting Signal Source and Scope

The model can be constructed either by selecting blocks from the Simulink Library Browser or by copying blocks from an existing model. Here, blocks

BUILDING A NEW SIMULINK MODEL

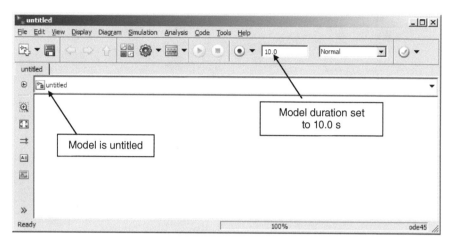

Figure 1.3 Simulink Model Blank Window.

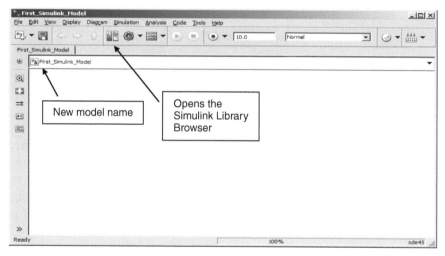

Figure 1.4 Simulink Model Window Renamed First_Simulink_Model.

will be copied from the library, which can be opened from the model window by clicking on the four-symbol icon on the toolbar shown in Figure 1.4. First, in the library window (Figure 1.2), click on **Sources** to open the window shown in Figure 1.5. With both the **First_Simulink_Model** window and the Simulink Library Browser window open, left-click on the **Sine Wave** icon and drag a copy into the model window. Alternatively, you can right-click on the icon in the library and select **Add to First_Simulink_Model**.

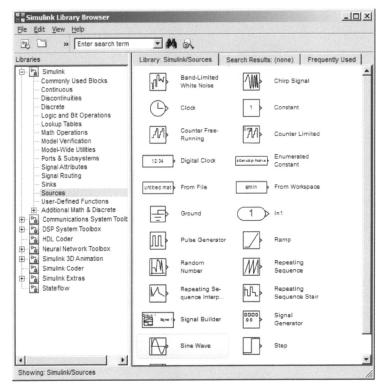

Figure 1.5 Simulink Library Browser with **Sources** Selected.

Next, add a scope to the model by returning to the Simulink Library Browser and clicking on **-Sinks**, selecting **scope**, and dragging a copy into the model window, now shown in Figure 1.6. In the figure, the **Sine Wave** block has been connected to the **Scope** by clicking on the arrowhead at the **Sine Wave** output and dragging a line to the corresponding arrowhead at the input to the **Scope**.

Another block, entitled **Model Info**, is shown in Figure 1.6 and is available in the Simulink library under **Simulink Model-Wide Utilities**. Dragging this block to the **First_Simulink_Model** and double clicking on this block opens a text box. This utility is very useful for conveniently displaying the parameters of each simulation model and identifying pertinent information about the model.

1.4.2 Setting the Source Block Parameters

In the model window, double-click on the **Sine Wave** icon; this opens an information window for the **Sine Wave** block, shown in Figure 1.7. In the

BUILDING A NEW SIMULINK MODEL

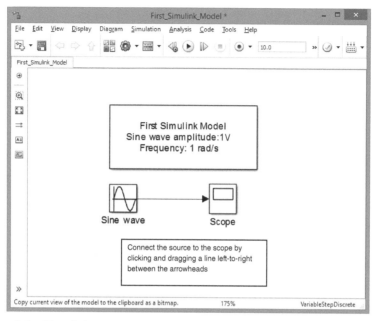

Figure 1.6 Simulink Model With Sine Wave Source, Scope, and Model Info Blocks.

window, the **Amplitude** and **Bias** of the sinusoidal source are selected as 1 and 0, respectively. The **Sine** type and **Time (t)** selections shown in the figure will be suitable for most simulations. The **Sine Wave** block parameters window can also be opened by selecting the block in the model window and right-clicking the mouse, which displays a list of options, and selecting **Block Parameters (Sin)**. This same pull-down menu provides options to manipulate and/or format the selected block, which the user will find helpful in structuring block diagrams in Simulink models.

The user will normally find options that can be selected for each block in addition to entering parameter values. As an example, in this block the user has a choice under **Sine** type to select either **Time-based** or **Sample-based** computation; under **Time (t)** the user can select **Use simulation time** or **Use external signal**. To provide a stream-lined introduction to model building, the presentation in this text will omit detailed discussions of many options. The Simulink documentation, accessible by clicking on the **Help** button, provides more extensive information on the optional selections.

1.4.3 Setting Scope Parameters

In the model window, double-click on the **Scope** icon, opening the **Scope** display, shown in Figure 1.8. At this point, the display is blank, since no

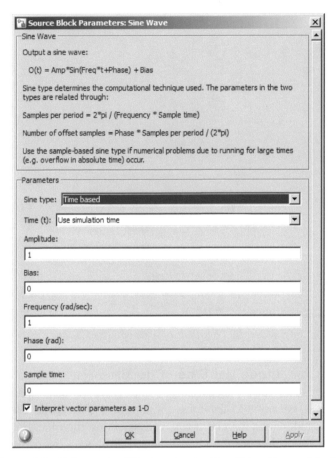

Figure 1.7 Information Window for the Sine Wave Source Block.

simulation has been started with this model. The gear-like icon on the toolbar opens the **Scope Parameters** window, which has three pages. On the **General** page, shown in Figure 1.9, the user is able to set the number of axes to be displayed in the **Scope** display, the simulation time range to be displayed, and to specify where **Tick labels** are to be applied in the display. Since the scope can be set to display multiple axes (for multiple inputs), the time ticks might be applied to all axes (select **all**), or to **none**, or to the **bottom axis only**.

The **History** page of the **Scope Parameters** window is shown in Figure 1.10. Here the user can specify the number of simulation data points to be displayed on the scope and can elect to have the data stored to **workspace**. On the **Style** page of the **Scope Parameters** window, shown in Figure 1.11, the user has various options as to colors, line styles and choice of markers to be used in the scope display.

EXECUTING THE SIMULINK MODEL 11

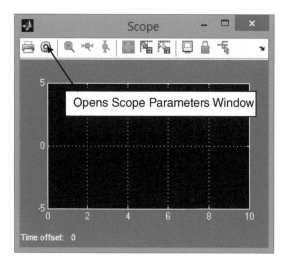

Figure 1.8 Scope Display.

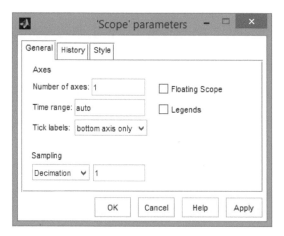

Figure 1.9 Scope Parameters Window: General Page.

1.5 EXECUTING THE SIMULINK MODEL

To execute the simple **Sine Wave** model, in the model window, click on the dark arrow button near the center of the toolbar, as shown in Figure 1.12. Double-clicking on the **Scope** block opens the **Scope** display window, shown in Figure 1.13. Examining the signal trace on the **Scope** display confirms the signal source settings made in Figure 1.7: **Amplitude = 1**, and **Frequency = 1 rad/s**.

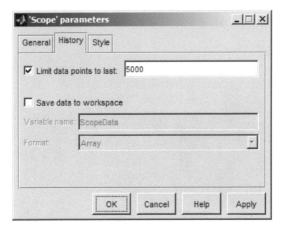

Figure 1.10 Scope Parameters Window: History Page.

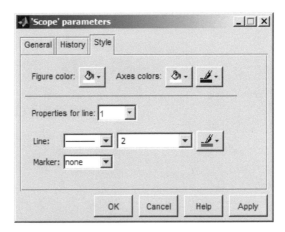

Figure 1.11 Scope Parameters Window: Style Page.

In this model, under the **Display** tab, selecting **Sample Time** and then all adds a data label **Cont** indicating that continuous time has been chosen, a direct result of choosing **Sine type** as **Time based** in Figure 1.7. The model colors are now changed, as seen in Figure 1.14, where the **Sine Wave** and **Scope** blocks are black and the **Model Info** block is magenta.

A **Sample Time Legend** is also displayed, as shown in Figure 1.15, indicating the block colors, the data type and the data values. In this case, the **Sine Wave** block is black, labeled **Cont** for continuous with a zero value; the **Model info** block is magenta with data value and type **Inf,** indicating that there is no time associated with this block, that is, it exists as long as the simulation is active.

EXECUTING THE SIMULINK MODEL

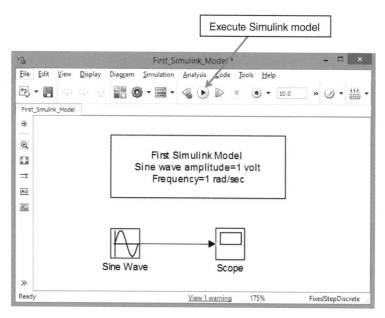

Figure 1.12   Executing the Simulink Model.

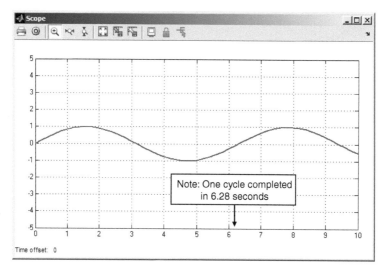

Figure 1.13   Scope Display After Executing the Sine Wave Model.

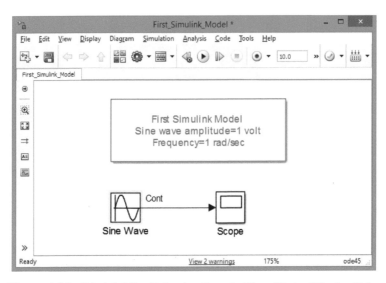

Figure 1.14 Model After Selecting Sample Time Under Display Tab.

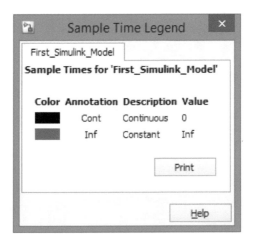

Figure 1.15 Sample Time Legend Indicating Block Characteristics.

1.6 RECONFIGURING THE SIGNAL BLOCK

At this point, another model change will be used to demonstrate an important feature of Simulink: Simulink blocks are designed to accept signals and parameter values as vector inputs. As an example, the time-based **Sine Wave** model in Figure 1.14 can be modified to generate two sinusoids instead of just one. A simple way of doing this is double-clicking on the

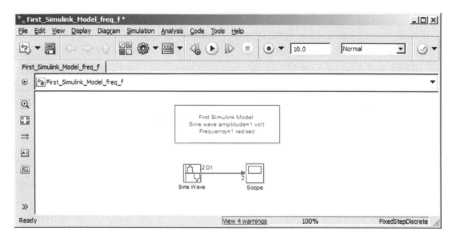

Figure 1.16 Sine Wave Model For Generating Sinusoids at 1 and 10 rad/s.

Sine Wave block, opening the block parameters window, and changing the **frequency (rad/s)** setting by inserting the two-element vector [1 10]. This configures the **Sine Wave** block to generate two sinousoids, one at 1 rad/s, the other at 10 rad/s. This change creates a new model, here renamed **First_Simulink_Model_freq_f**, shown in Figure 1.16. Note in the model window that the **Sine Wave** output is labeled **2D1** and the input to the **Scope** is labeled 2, both resulting from setting the frequency to the vector value [1 10].

After running the simulation, the **Scope** will display the two sampled sinusoids, as seen in Figure 1.17. To generate more frequencies with the **Sine Wave** block, simply add more terms in the vector input for the **frequency (rad/s)** cell in the parameters window.

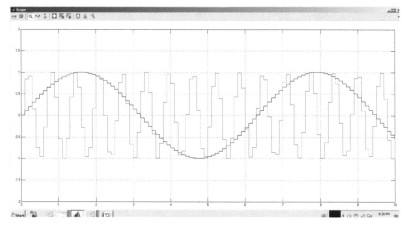

Figure 1.17 Scope Display for First Simulink Model with Frequency Set to [1 10].

Another way of configuring the **Sine Wave** block to generate the two sinusoids is to specify the **frequency (rad/s)** as **f** in the block parameters window, and define the variable **f** by inserting the statement **f = [1 10]** into the MATLAB Command Window. A third way of reconfiguring the **Sine Wave** block is discussed in Section 1.9, where the use of **Model Explorer** is demonstrated.

1.7 SAMPLE-BASED SIGNALS

In Figure 1.18, the simulation model is changed by selecting **Sample based** in the **Source Block Parameters: Sine Wave** window; the **Sample time** is set to 0.1 (where the units are seconds) and the number of **samples**

Figure 1.18 Sample Based Selection in Sine Wave Block.

SAMPLE-BASED SIGNALS

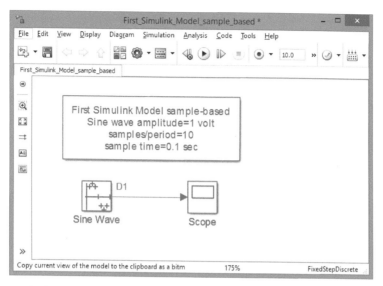

Figure 1.19 First Simulink Model Modified for Sample Based computation.

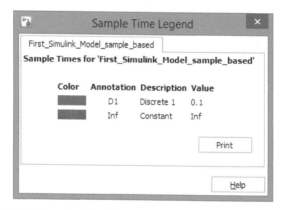

Figure 1.20 Sample Time Legend for Sample Based Computation.

per period is set to 10. Figures 1.19 and 1.20 display, respectively, the sample-based model and the **Sample Time Legend** resulting from the change. The color red in the blocks and data line indicates that the output from the **Sine Wave** block is a discrete vector with 0.1 s time steps and the magenta color shows **inf** corresponding to an infinite associated time.

Upon running the simulation, the scope output, shown in Figure 1.21, displays the resulting sample values as a function of time (Note: labeling of axes is discussed in a later section.)

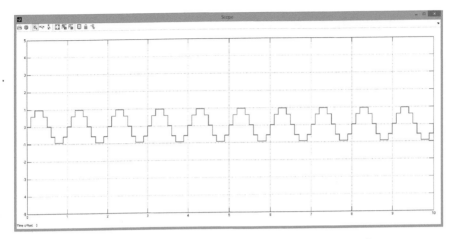

Figure 1.21 Scope Display for Sample-Based Computation.

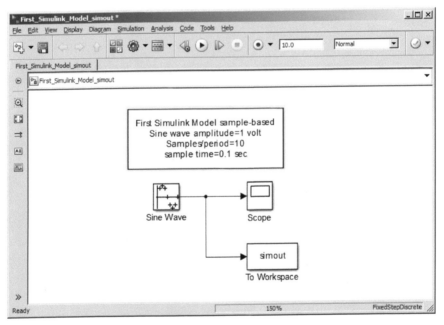

Figure 1.22 Sending Simulation Data to Workspace.

1.8 SENDING DATA TO WORKSPACE

In Figure 1.22, **First Simulink Model-sample-based** has been augmented by adding the **To Workspace** block from **Simulink Sinks** while

USING MODEL EXPLORER 19

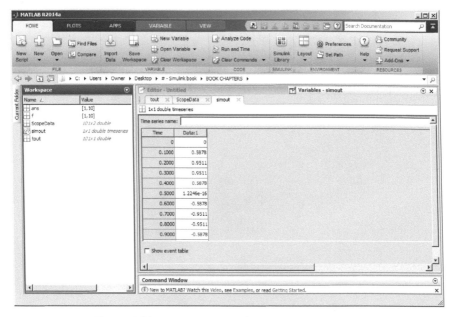

Figure 1.23   Partial Output for the Variable simout.

retaining sample based computation; the model has been renamed by saving it as **First_Simulink_Model_simout**. The **To Workspace** block causes the output data from the **Sine Wave** block to be saved and examined for subsequent use such as plotting. The output block is labeled here as **simout**, a label that can be changed after first double clicking on the block. After running the model, the simulation results are retrieved by clicking on the variable **simout** in the MATLAB Workspace window. Figure 1.23 shows partial output data for the simulation, where both the time increments and sampled sine wave data values are displayed.

1.9   USING MODEL EXPLORER

Model Explorer is a tool available to provide the user with the ability to view, modify or add elements in the Simulink model and **workspace variable**s. Refer to the MathWorks documentation for further information on the use of **Model Explor**er.

To open the **Model Explorer**, select **Model Explorer** under the **View** tab in the Simulink model window. Figure 1.24 shows the window displayed

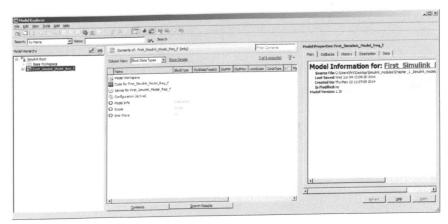

Figure 1.24 Model Explorer Window for First_Simulink_Model_freq_f.

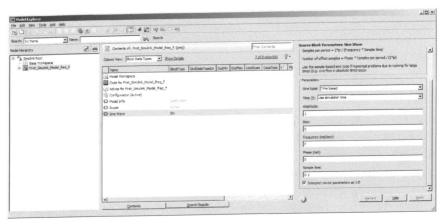

Figure 1.25 Model Explorer Showing Sine Wave Block Information.

after performing this selection from the **First_Simulink_Model_freq_f** model window of Figure 1.16. The menu in the center panel lists all the blocks in the selected model.

By clicking on any model block in the list, the user can display information about that block. For example, Figure 1.25 displays information for the **Sine Wave** block, where the frequency parameter **f** is seen.

This provides another avenue for reconfiguring the **Sine Wave** block to generate multiple frequencies, discussed earlier. That is, the **First_Simulink_Model_freq_f** could have been modified in the **Model Explorer** window by setting **frequency (rad/s)** to **[1 10]**. To accomplish the task of setting the frequency parameter f, open the **Callbacks** tab in the

ADDING LABELS TO FIGURES

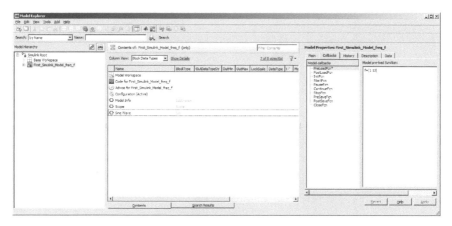

Figure 1.26 Frequency Parameter f set in Model Explorer Callbacks Preloadfcn*.

right-hand panel of the **Model Explorer** window, and in the **PreLoadFcn***
enter **f = [1 10]** as seen in Figure 1.26.

1.10 ADDING LABELS TO FIGURES

The Simulink **scope** block does not support the manipulation of graphics properties in **scope** displays. To add labels to Figure 1.17 the following snippet of Matlab code is used[3]:

```
shh = get(0,'ShowHiddenHandles');
set(0,'ShowHiddenHandles','On');
set(gcf,'menubar','figure');
set(gcf,'CloseRequestFcn','closereq');
set(gcf,'DefaultLineClipping','Off');
set(0,'ShowHiddenHandles',shh);
```

Entering this snippet in the the MATLAB Command Window displays an **Edit** tab in the figure. Selecting **Figure Properties** from the **Edit** tab produces the window shown in Figure 1.27. Changes can now be made to colors, style, axis properties, and labels.

After adding labels, inserting a title, and changing colors, the revised figure is displayed in Figure 1.28.

---

[3]This snippet of MATLAB code was provided by the MathWorks technical staff.

22 GETTING STARTED WITH SIMULINK

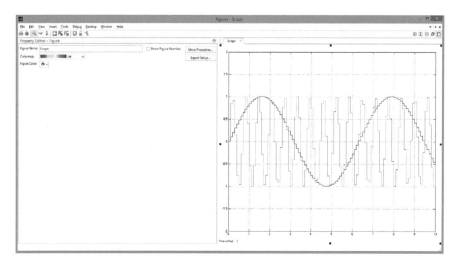

Figure 1.27 Figure Properties Window.

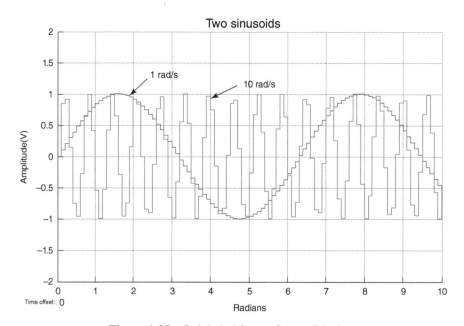

Figure 1.28 Relabeled Scope Output Display.

1.11 SELECTING MODEL CONFIGURATION PARAMETERS

In the Simulink model window, pulling down the **Simulation** tab and selecting **Model Configuration Parameters** opens a window where the user

SELECTING MODEL CONFIGURATION PARAMETERS

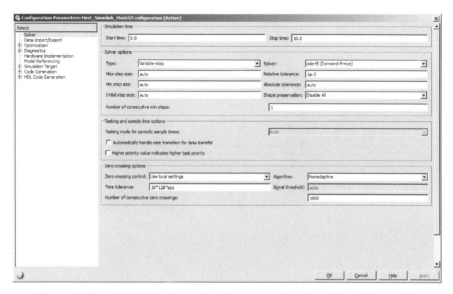

Figure 1.29 Model Configuration Parameters for First Simulink Model.

can specify the simulation start and stop time and choose the solver for the simulation[4]. Figure 1.29 shows the **Model Configuration Parameters** for **First_Simulink_Model** where the **Type** of solver is set to **Variable-step** and the **Solver** is selected as **ode-45(Dormand-Prince**), which, in general is the best first choice as a solver for most Simulink models. By clicking on the **Help** button in the **Model Configuration Parameters** window, the MathWorks documentation describes several aspects of this window including:

- Solver choices
- Simulation and clock time are not the same
- Fixed-step and variable-step size
- Shortened simulation time with variable-step solver

For both fixed-step and variable-step solvers, the next simulation time is the sum of the current simulation time and the step size. Using a fixed-step solver, the step size remains constant throughout the simulation whereas use of a variable-step solver allows the step size to vary from step to step in accordance with the specified error tolerance.

[4]Simulink provides a variety of solvers, each appropriate for running a particular type of simulation model. Detailed discussions of the various solvers can be found in Dabney and Harmon, and in Jamshidi, Farzad and Pedar, cited in Appendix B.

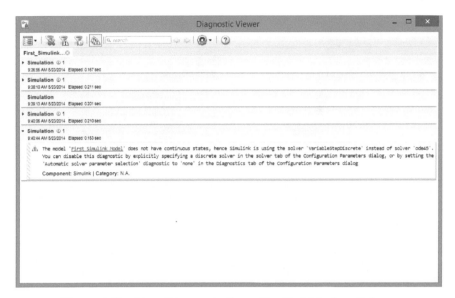

Figure 1.30 Error Diagnostic Caused by Variable-Step Solver.

If the **First_Simulink_Model** is modified where the **Sine Wave** block is chosen to have a **Sample-based** Sine type and a 0.1 s **Sample time** is entered, the model execution will produce a warning message seen at the bottom of the Simulink model. The error diagnostic explains the problem as seen in Figure 1.30.[5]

In Figure 1.31, the **Solver** is changed to **Fixed-step** in the **Solver type** tab and discrete in the **Solver** tab so that no error message is generated.

1.12 SUMMARY DISCUSSION

This chapter has presented a brief introduction to Simulink by demonstrating the basic steps to be taken in constructing a simple simulation model. The MathWorks documentation provides a comprehensive treatment of each block and available tools. The remaining chapters focus on the use of Simulink for modeling digital communications systems, without delving deeply into aspects of the full Simulink capability. Other references that the user is likely to find helpful are listed in Appendix B.

---

[5] Error diagnostics are often generated when a simulation is first developed. Care must be taken by the user in that the error message identifies a problem but the solution may reside in a block other than the highlighted block. The Simulink toolbar has a check button that can invoke the Model Advisor to help the user correct the problem.

PROBLEMS

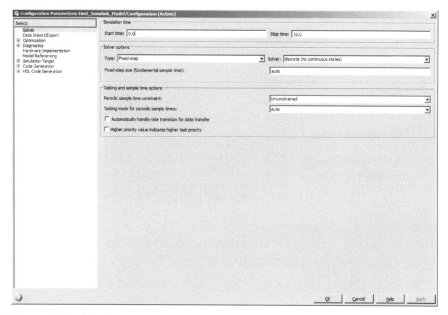

Figure 1.31 Model Configuration Parameters with Solver Changed to Fixed-Step Discrete.

PROBLEMS

1.1 Modify the Simulink model in Figure 1.12 to produce two sinusoidal waves with the following parameters:

frequency = 1 rad/s for both waves
amplitude = 1 V for both waves
phase = 0 for one wave and $\pi/2$ for the second wave
sample time = 0.01 s

a. Show the Simulink Model and include an information block.

b. Display each wave on a separate trace in the scope and label all axes.

Hint: Find the demux block in the Simulink library.

1.2 Let $x(t) = \frac{4}{\pi}\left[\sin(t) + \frac{1}{3}\sin(3t) + \frac{1}{5}\sin(5t)\right]$.

a. Develop a Simulink model for $x(t)$ with an included information block. Assume a 10 s simulation time.

b. Display $x(t)$ in a scope over the range 0 to 2π with labels.

c. Modify the Simulink model obtained in part a by overlaying a square wave that is +1 between 0 and π and −1 from π to 2π and repeats thereafter.

d. Display the overlay result in a scope over the range 0 to 2π with labels.

Note that $x(t)$ represents the first three terms of the Fourier series of the square wave.

1.3 Amplitude Modulation (AM) with a tone modulator having a unity modulation index is expressed as

$$x(t) = (1 + \cos(t)) \cos(20t)$$

a. Develop a Simulink model for $x(t)$ with an included information block. Use a 10 s simulation time and Goto and From routing blocks from Signal Routing to simplify the model.

b. Display $x(t)$ and $\cos(t)$ on a scope with labeled axes.

c. From the Simulink library, add an AM modulation block to the simulation and form the difference between $x(t)$ and the output of the AM library block.

d. Display $x(t)$, $\cos(t)$, the AM block output and the difference on a scope with four traces. Insert x axis title on bottom trace only; do not label y-axis but add a title to each plot.

1.4 Develop a Simulink model with a sine wave input that feeds both a double-sideband (DSB) AM block and a quantizer followed by a DSB AM block. Assume a 2 s simulation and sine wave block parameters as follows:

Sine wave amplitude = 2, Frequency = 20π rad/s, Sample time = 0.001 s.

For the DSB AM block, assume that the parameters are:

input signal offset = 1, carrier frequency = 100, initial phase = 0

a. Show the model with an included information block.

b. Assume the quantization interval = 0.5 and display the following signals in a scope with 4 traces:

sine wave output, DSB AM output, quantizer/DSB AM output, difference between the DSB AM output and quantizer/DSB AM output.

Provide titles for each trace and label only the x-axis.

c. Repeat part a with a quantization interval = 0.05.

2

SINUSOIDAL SIMULINK MODEL

2.1 A FIRST SIMULINK MODEL

This chapter continues focusing on the development of a simple Simulink model based on a sinusoidal signal. Specific topics include:

- Simulink model of a sine wave
- Spectrum of a sine wave

2.2 SIMULINK MODEL OF SINE WAVE

This section continues focusing on the development of the Simulink model of a sine wave presented in Chapter 1. A general representation of a sine wave with amplitude A, carrier frequency f_c and phase φ is

$$x(t) = A \sin(2\pi f_c t + \varphi)$$

A Simulink model for displaying and analyzing its characteristics is shown in Figure 2.1.

Modeling of Digital Communication Systems Using SIMULINK®, First Edition.
Arthur A. Giordano and Allen H. Levesque.
© 2015 John Wiley & Sons, Inc. Published 2015 by John Wiley & Sons, Inc.
Companion Website: www.wiley.com/go/simulink

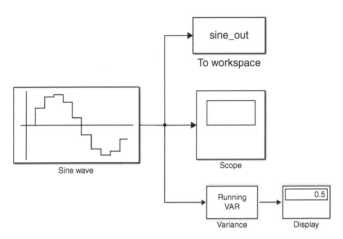

Figure 2.1 Simulink Model for Sine Wave.

The Simulink model parameters for this example are specified as follows:

Model Parameters for Sine Wave

- Sample time = 0.05 s
- Run time = 2 s
- Parameters: $A = 1, f_c = 1, \varphi = 0$

The Source Block Parameters: Sine Wave are provided in Figure 2.2.

The parameters for the To Workspace block, sine_out, are shown in Figure 2.3.

Figure 2.3 shows that the Save format entry is selected to be a Timeseries. This selection enables the plot choice in the MATLAB main window to be utilized by clicking on sine_out in the workspace. As a result, the plot from the scope is produced as shown in Figure 2.4. Selecting Edit Figure Properties from the Edit menu in the plot allows the user to change the figure scales, color, line width, etc.

The theoretical average power of $x(t)$ is $A^2/2 = 0.5$ and is in agreement with the output estimated by the running variance block.

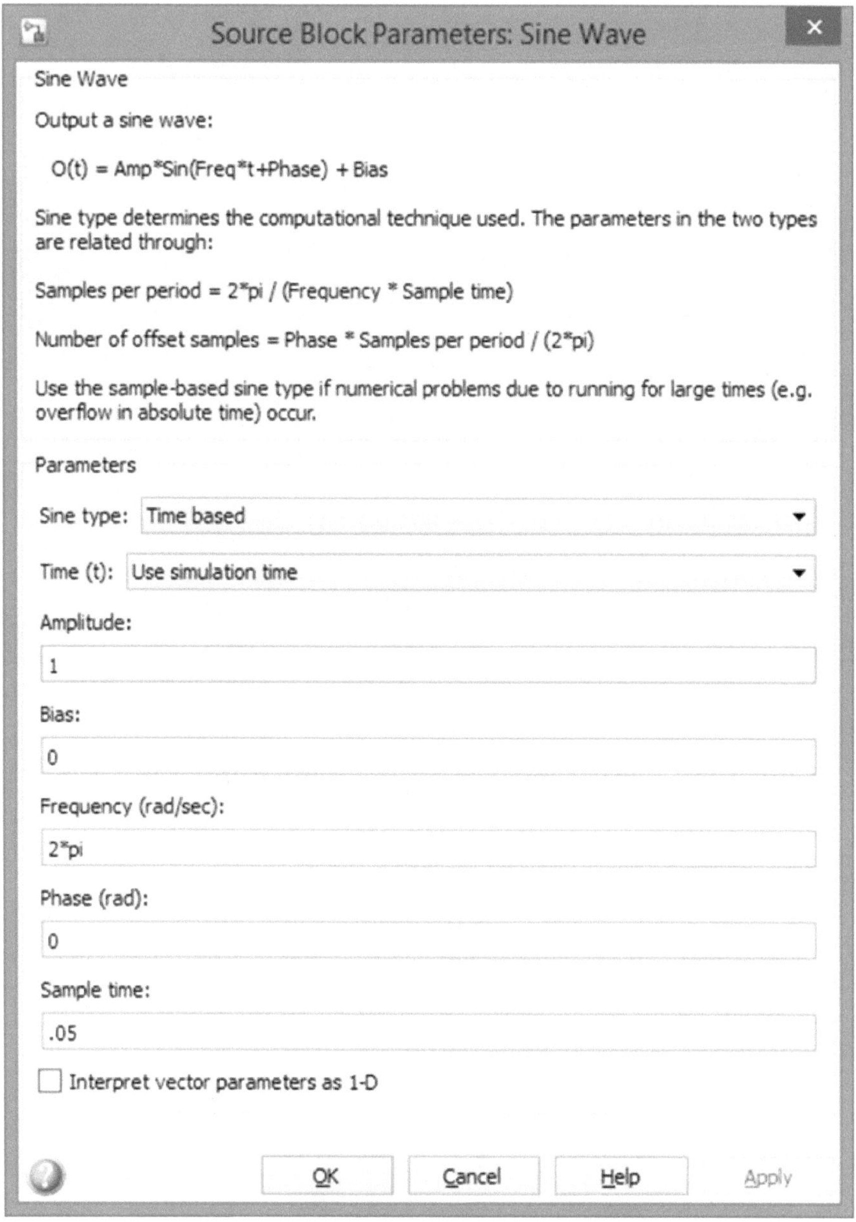

SIMULINK MODEL OF SINE WAVE

Source Block Parameters: Sine Wave

Sine Wave

Output a sine wave:

O(t) = Amp*Sin(Freq*t+Phase) + Bias

Sine type determines the computational technique used. The parameters in the two types are related through:

Samples per period = 2*pi / (Frequency * Sample time)

Number of offset samples = Phase * Samples per period / (2*pi)

Use the sample-based sine type if numerical problems due to running for large times (e.g. overflow in absolute time) occur.

Parameters

Sine type: Time based

Time (t): Use simulation time

Amplitude:
1

Bias:
0

Frequency (rad/sec):
2*pi

Phase (rad):
0

Sample time:
.05

☐ Interpret vector parameters as 1-D

OK   Cancel   Help   Apply

Figure 2.2 User Inputs for Sine Wave Block.

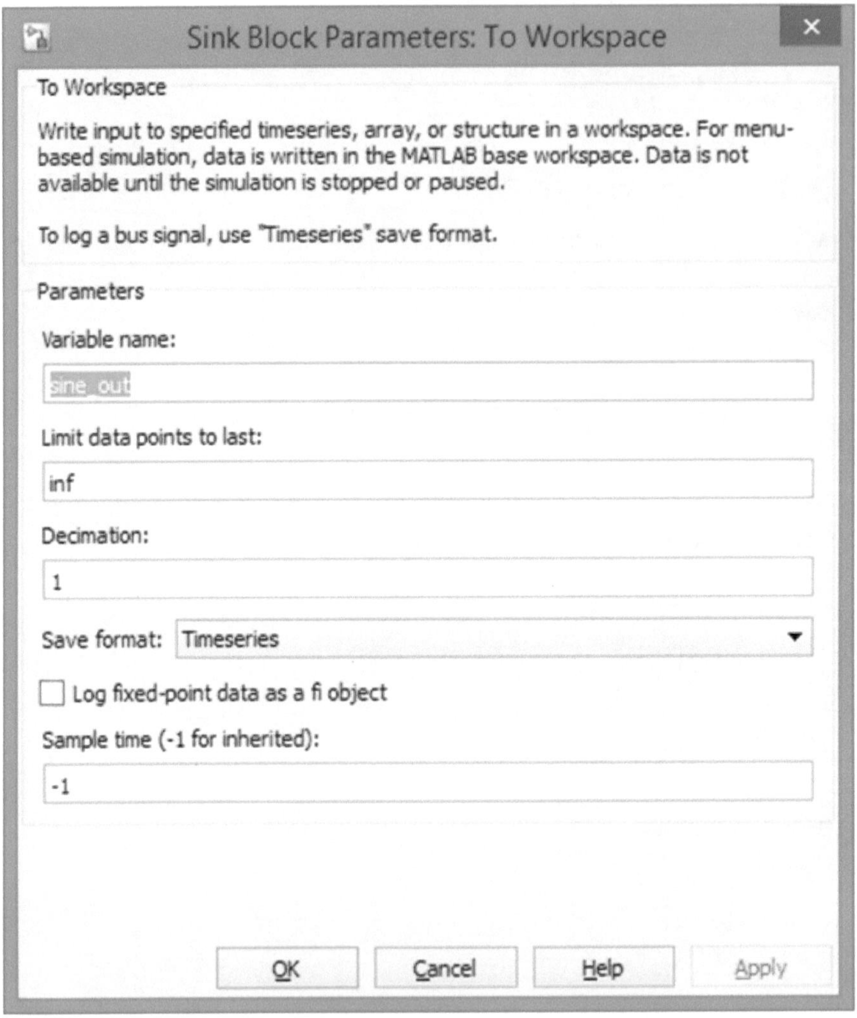

Figure 2.3 Sink Block Parameters: To Workspace.

Figure 2.5 shows the variables in the Workspace listed as data, sine_out and tout where a partial series of the values is displayed. The variable tout is the sample values of time displayed every 0.05 s; the variables data and sine_out both list the amplitude values of the sine wave but in different formats.

Table 2.1 summarizes partial results where column 1 is the time value, column 2 is the value from the simulation, and column 3 is the computed analytic value demonstrating good agreement with the simulation.

SIMULINK MODEL OF SINE WAVE

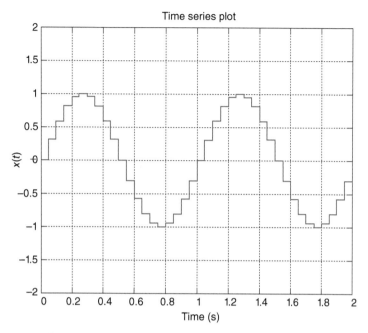

Figure 2.4 Plot of Sine Wave, $x(t)$ sampled at 0.05 s.

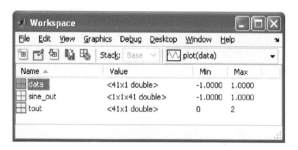

Figure 2.5 Workspace Variables for Simulink Sine Wave.

TABLE 2.1 Comparison of Simulation and Analytic Values

Time (nT_s) $T_s = 0.05$ s	Simulink Value	Sin $(2\pi nT_s)$
0	0	0
0.05	0.3090	0.3090
0.10	0.5878	0.5878
0.15	0.8090	0.8090
0.20	0.9511	0.9511
0.25	1.0000	1.0000
0.30	0.9511	0.9511
0.35	0.8090	0.8090
0.40	0.5878	0.5878
0.45	0.3090	0.3090
0.50	4.44089209850063 e-16	0

Next a simple change is made where the samples of the sine wave are taken more frequently at 0.01 s as shown in Figure 2.6. Figure 2.7 depicts the results given this change.

Next a change is made to include a phase shift in the sine wave block as displayed in Figure 2.8

Figure 2.9 shows the results with 0 and a $\pi/2$ phase shift where the bottom figure is clearly $\cos(2\pi t)$. Nyquist sampling requires only 2 samples/cycle but the figures are easier to read with more samples.

2.3 SPECTRUM OF A SINE WAVE

This section now examines the spectrum of a sine wave using Simulink library blocks. The theoretical spectrum, $X(f)$, of $x(t) = A \cos(2\pi f_c t)$ is

$$X(f) = \frac{A}{2}\delta(f - f_c) + \frac{A}{2}\delta(f + f_c)$$

and its power spectral density (PSD), $S_x(f)$, is

$$S_x(f) = \frac{A^2}{4}\delta(f - f_c) + \frac{A^2}{4}\delta(f + f_c)$$

For the example presented here, let $A = 1$ and $f_c = 100$ Hz. The peak value of the spectrum is then $1/2$ or -3 dB. The average power is $A^2/2 = 0.5$.

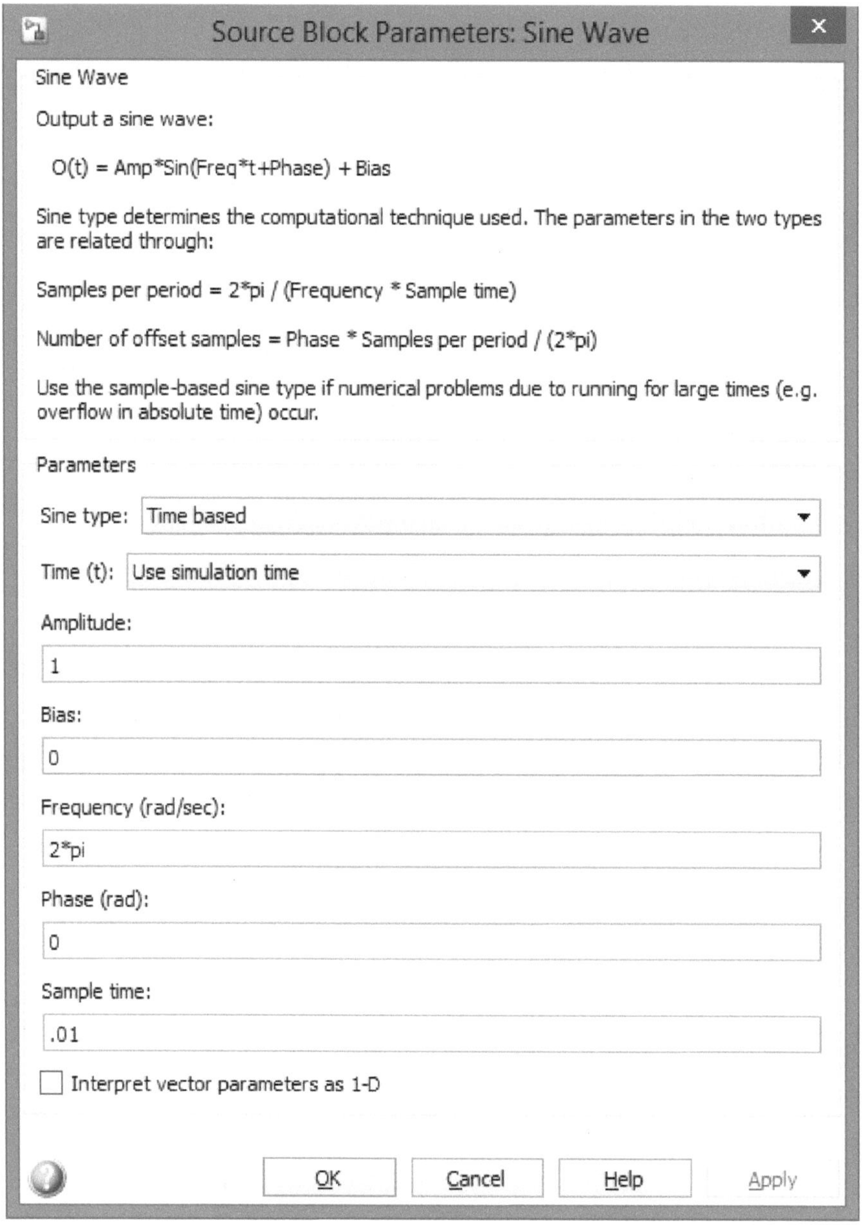

SPECTRUM OF A SINE WAVE

Source Block Parameters: Sine Wave

Sine Wave

Output a sine wave:

O(t) = Amp*Sin(Freq*t+Phase) + Bias

Sine type determines the computational technique used. The parameters in the two types are related through:

Samples per period = 2*pi / (Frequency * Sample time)

Number of offset samples = Phase * Samples per period / (2*pi)

Use the sample-based sine type if numerical problems due to running for large times (e.g. overflow in absolute time) occur.

Parameters

Sine type: Time based

Time (t): Use simulation time

Amplitude:
1

Bias:
0

Frequency (rad/sec):
2*pi

Phase (rad):
0

Sample time:
.01

☐ Interpret vector parameters as 1-D

[OK] [Cancel] [Help] [Apply]

Figure 2.6 Sine Wave Sampled at 0.01 s.

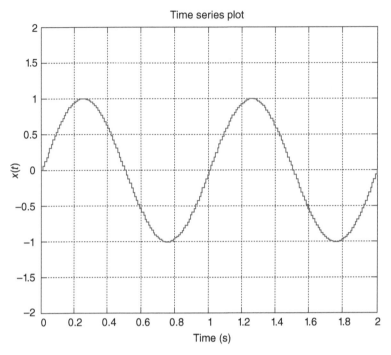

Figure 2.7 Plot of Sine Wave, $x(t)$ sampled at 0.01 s.

The peak of the power spectrum is $A^2/4$ and is expressed in dBW as $10\log(1/4) = -6\,\text{dBW}$.

A Simulink model for determining the spectrum and power spectrum of a sinusoidal signal is shown in Figure 2.10

The Simulink model parameters for this example are specified as follows:

Model Parameters for Cosine Wave $x(t) = A\cos(2\pi 100t)$

- Sample time = 0.001 s
- Run time = 10 s
- Parameters: $A = 1$, $f_c = 100 \times 2\pi$ rad/s, phase = $\pi/2$ rad
- FFT length bsize = 2048; buffer overlap = 0; scale = 1/bsize = 1/2048
- FFT: periodogram method
- Spectrum analyzer: Hann window

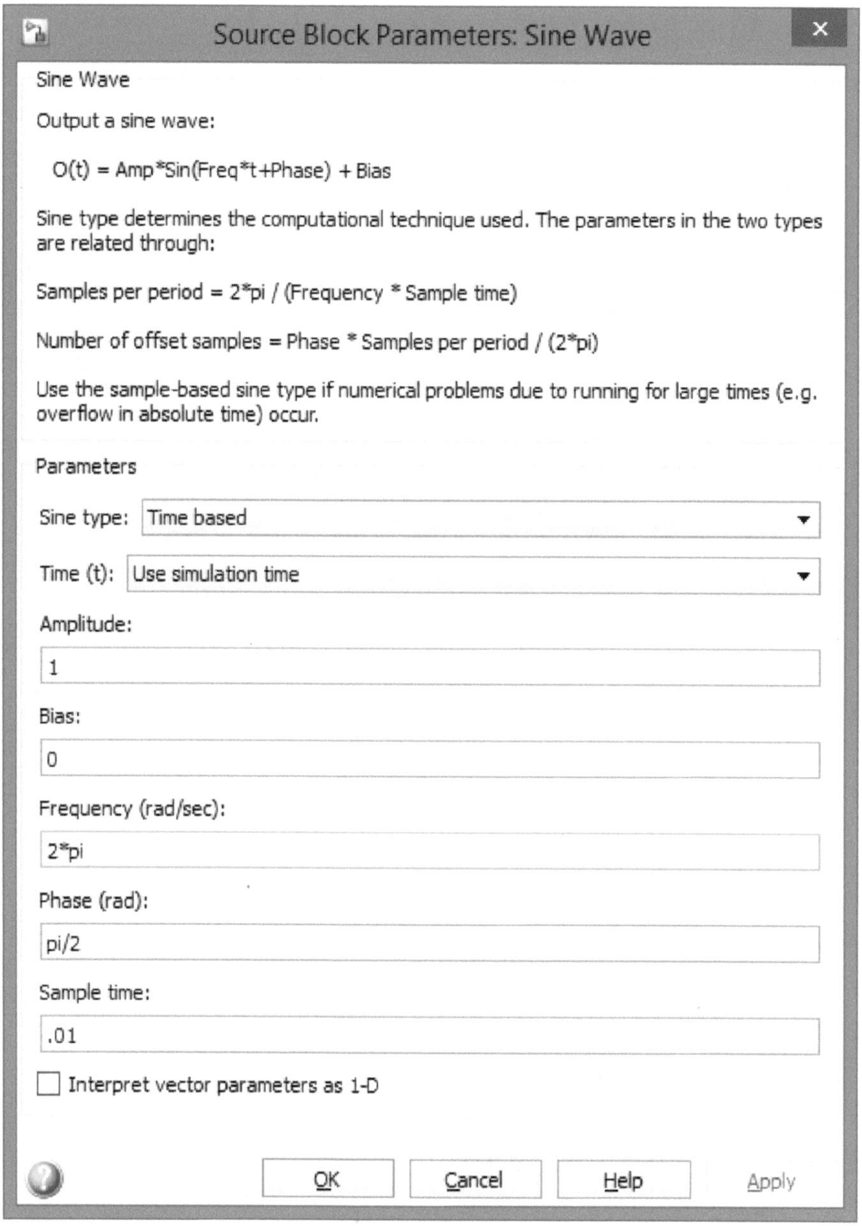

SPECTRUM OF A SINE WAVE

Source Block Parameters: Sine Wave

Sine Wave

Output a sine wave:

O(t) = Amp*Sin(Freq*t+Phase) + Bias

Sine type determines the computational technique used. The parameters in the two types are related through:

Samples per period = 2*pi / (Frequency * Sample time)

Number of offset samples = Phase * Samples per period / (2*pi)

Use the sample-based sine type if numerical problems due to running for large times (e.g. overflow in absolute time) occur.

Parameters

Sine type: Time based

Time (t): Use simulation time

Amplitude:
1

Bias:
0

Frequency (rad/sec):
2*pi

Phase (rad):
pi/2

Sample time:
.01

☐ Interpret vector parameters as 1-D

OK   Cancel   Help   Apply

Figure 2.8 Sine Wave Sampled at 0.01 s with a $\pi/2$ phase shift.

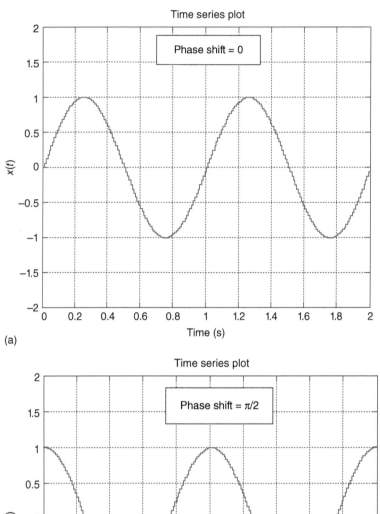

Figure 2.9 $x(t) = \sin(2\pi t)$ (a); $x(t) = \sin(2\pi t + \pi/2)$ (b).

SPECTRUM OF A SINE WAVE

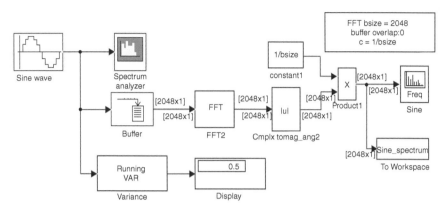

Figure 2.10 Simulink Model for Determining the Spectrum & Power Spectrum of a Sinusoidal Wave.

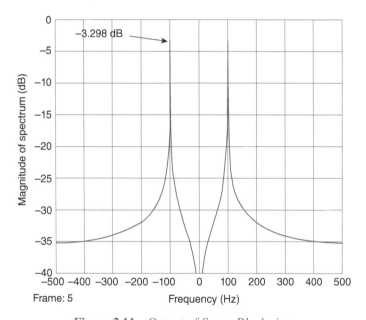

Figure 2.11 Output of Scope Block sine.

Figure 2.11 displays the output of the scope block labeled sine computed by means of the periodogram method[1]. In this figure, it can be observed that the

---

[1] Schonhoff, T.A. and A.A. Giordano, Detection and Estimation Theory, Pearson Prentice Hall, 2006, Chapter 19.

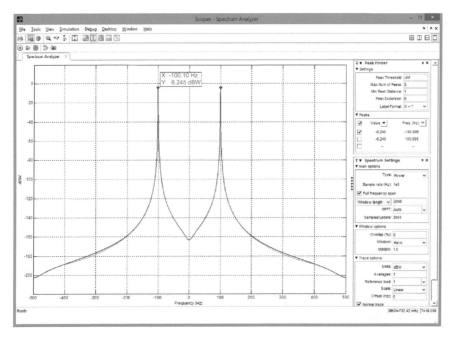

Figure 2.12 Output of Spectrum Analyzer with 2048 Point FFT and Hann Window.

magnitude of the spectrum $|X(f)|$ occurs at frequencies at $+100$ and -100 Hz. From the MATLAB Workspace, it can be observed that the peak value is 0.4680 or -3.298 dB. This peak value is lower than the theoretical -3 dB value.

Data windowing controls the width of the main spectral lobe and sidelobe leakage. The rectangular window used to produce Figure 2.11 provides good spectral resolution but increases the sidelobes. Figure 2.12 shows the spectrum analyzer output where a 2048 length FFT with no overlap is computed using a Hann window. In this figure, the frequencies are located at ± 100 Hz with peaks that are approximately -6 dBW with sidelobes that are much lower than those in Figure 2.11.

Figure 2.13 is an edited version of the spectrum analyzer output where the vertical scale is adjusted to correspond with the vertical scale in Figure 2.12. The peak of the spectrum is about -6.245 dBW, which is close to the expected theoretical peak value.

Figure 2.14 shows the spectrum analyzer output where a 2048 length FFT with no overlap is computed using a Flat Top window. This window provided a closer estimate of the spectrum peak in comparison with the theoretical value.

SPECTRUM OF A SINE WAVE

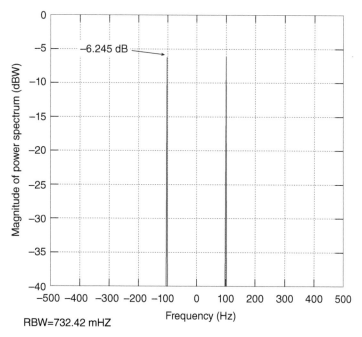

Figure 2.13 Output of Spectrum Analyzer with 2048 Point FFT and Hann Window and Adjusted Scale.

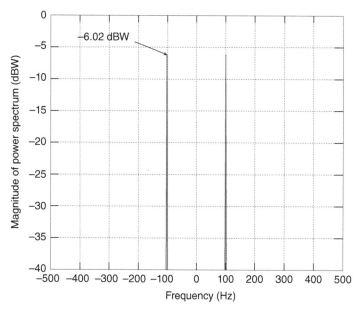

Figure 2.14 Output of Spectrum Analyzer with 2048 Point FFT and Flat Top Window.

2.4 SUMMARY DISCUSSION

This chapter has presented rudimentary Simulink models to allow the user to gain familiarity with the tools. Since MATLAB underlies all Simulink computations, the user should not be surprised at the numerical agreement between MATLAB and Simulink computations. The spectrum calculations using an FFT have demonstrated the need to utilize a window function to control the main lobe width and sidelobe leakage. The length of the FFT, the buffer overlap, the number of averages and the window selection directly determine the accuracy of the spectrum.

PROBLEMS

2.1 Find the average power of a sinusoid with the following parameters:

frequency = 1 rad/s
amplitude = 0.5 V
sample time = 0.01 s

a. Show the Simulink Model using the running Variance block and include an information block. Assume a 100 s simulation time.

b. Display the output in the scope and label all axes.

c. Show the results in a Display block and compare with the theoretical value.

2.2 In the Simulink model shown in Figure 2.1 enter the vector [1 4] for amplitude and [2*pi 20*pi] for frequency. Use a 0.001 s sample time and a simulation time of 2 s.

a. Display the results in the scope.

b. Show the Simulink model with the simulated average powers and compare the result with the theoretical powers.

c. Develop a model that allows each sinusoid to be displayed separately and label all axes in the scope output.

2.3 Develop a sample-based Simulink model for a sine block connected to a scope by:

1. Selecting the sine block from the Simulink section of the Simulink library

2. Selecting the sine block from the DSP sources in the Simulink library

   Assume the phase is zero, the sample time is 0.05 s and the frequency is 1 Hz.

   a. Compute the power for both signals
   b. Plot the scope output and compare the results

2.4 Find the average power in the square of a sine wave. Assume the following parameters:

frequency = 2π rad/s
amplitude = 1 V
sample time = 0.01 s

a. Show the Simulink Model with an information block. For a 10 s simulation time compute the power of the squared sine wave using the running Variance and running RMS blocks and show the results in Display blocks

b. Display the output in the scope and label all axes

c. Repeat Part a. using 100 s simulation

d. Compare the results in Parts a. and c. with theoretical values

2.5 In the Simulink model shown in Figure 2.1 insert an AWGN block at the output of the sine block and change the sample time to 0.01 s. Modify the scope to display the sine output and the output of the AWGN block

a. Display the revised Simulink model. Plot the results assuming the SNR = 0 dB in the AWGN block and report the signal power from the output of the AWGN block

b. Change the run time to 100 s and report the results of the signal power from the output of the AWGN block and explain the difference from Part a.

c. Change the SNR in the AWGN block to be 10 dB. Explain the resulting change in the power from the output of the AWGN block

2.6 The autocorrelation of $x(t) = A\cos(2\pi f_c t)$ is known to be $R(\tau) = \frac{A^2}{2} \cos(2\pi f_c \tau)$. Assume $A = 1$ and $f_c = 100$ Hz. Using the Simulink blocks $|FFT|^2$, IFFT and the Sine Wave block and associated parameters shown in Figure 2.10, develop a Simulink model to determine the autocorrelation and compare the result with the theoretical expression.

2.7 Modify the Simulink model shown in Figure 2.10 as follows:

a. Insert a 256 buffer overlap while retaining the 2048 FFT length. Compute the spectrum magnitude.

b. With no buffer overlap reduce the FFT length to 256. Compute the spectrum magnitude and the power spectrum magnitude using the spectrum analyzer.

2.8 Develop a Simulink model to determine the power spectrum of a vector of 4 sinusoids all having a unity amplitude and frequencies 100, 200, 300, and 400 Hz. Use the DSP sine wave block as a source and compute the spectrum using 10 spectral averages.

3

DIGITAL COMMUNICATIONS BER PERFORMANCE IN AWGN (BPSK AND QPSK)

3.1 BPSK AND QPSK ERROR RATE PERFORMANCE IN AWGN

This chapter introduces several topics in Simulink utilizing binary Phase-Shift Keying (BPSK) and quaternary Phase-Shift Keying (QPSK) modulations. Specifically these topics include:

- Construction of a Simulink model in simple steps
- Available menu choices
- Description of inputs and outputs in simulations
- Sample and frame-based computations
- Complex signal representations
- BPSK bit error rate (BER) simulated and theoretical results
- Comparison of additive white Gaussian noise (AWGN) and Gauss Simulink library blocks
- QPSK symbol error rate performance
- Fixed point error rate performance

Modeling of Digital Communication Systems Using SIMULINK®, First Edition.
Arthur A. Giordano and Allen H. Levesque.
© 2015 John Wiley & Sons, Inc. Published 2015 by John Wiley & Sons, Inc.
Companion Website: www.wiley.com/go/simulink

3.2 CONSTRUCTION OF A SIMULINK MODEL IN SIMPLE STEPS

The BPSK simulation for computing the bit error rate (BER) will be developed in a sequence of steps as follows:

- Begin with the simplest Simulink BPSK model in a back to back configuration with no additive noise and verify proper operation (step 1)
- Add an AWGN block with very high E_b/N_o so that additive noise does not introduce errors and verify proper operation (step 2)
- Add scopes at the output of the BPSK modulator and input to the BPSK demodulator and observe multiple errors when $E_b/N_o = -10\,\text{dB}$ (step 3)
- Add an error rate calculation block and an associated display to produce an estimate of the BER at a specified E_b/N_o value (step 4)
- Add an information block and display the signal and port data types (step 5)

Figure 3.1 displays a Simulink model that will be modified later to compute the BPSK BER. The blocks include a Random Integer source, a BPSK modulator, a BPSK demodulator and several output blocks used to transfer data to the MATLAB Workspace.

The menu selection for the Random Integer source block is shown in Figure 3.2. In the menu, binary numbers are selected with $M=2$ and the initial random seed $= 37$.[1] This is a sample-based computation where individual samples are issued in 1 s intervals. The outputs of this block are random double precision numbers that are either 0 or 1.

The BPSK modulator menu is shown in Figure 3.3 where a zero phase angle is selected as displayed in Figure 3.4 by clicking on the View Constellation button. A corresponding phase offset occurs in the BPSK demodulator. If the phase offsets angles between the modulator and demodulator do not agree, then errors will be made. Implementation considerations may force such a disagreement and the degradation experienced can then be determined.

Pressing the data tab reveals that the output is also double precision as shown in Figure 3.5. The output sequence of the BPSK modulator consists of complex-valued double precision numbers with real values equal to $+1$ or -1.

---

[1] Ordinarily, a prime number is selected as a seed.

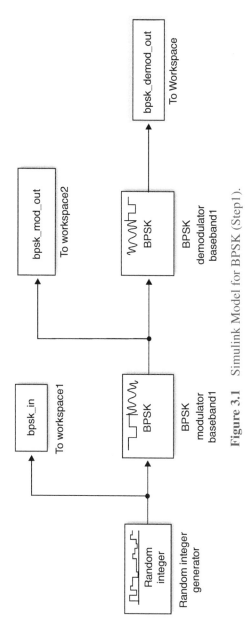

Figure 3.1 Simulink Model for BPSK (Step1).

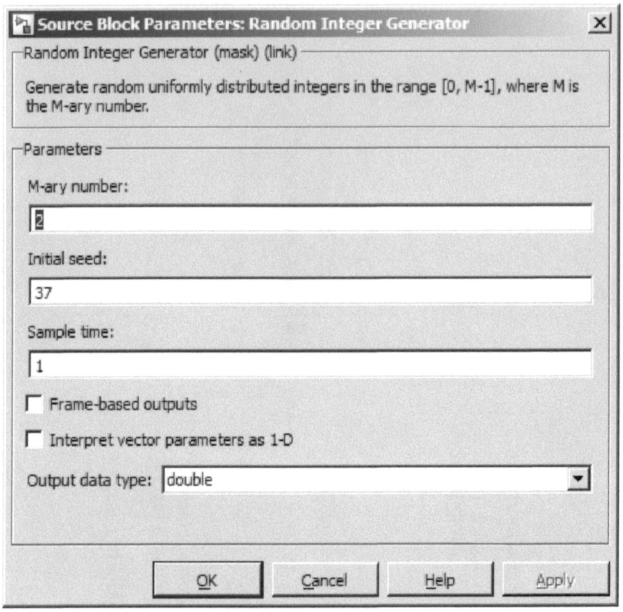

Figure 3.2 Random Source Block for Input Parameter Selection.

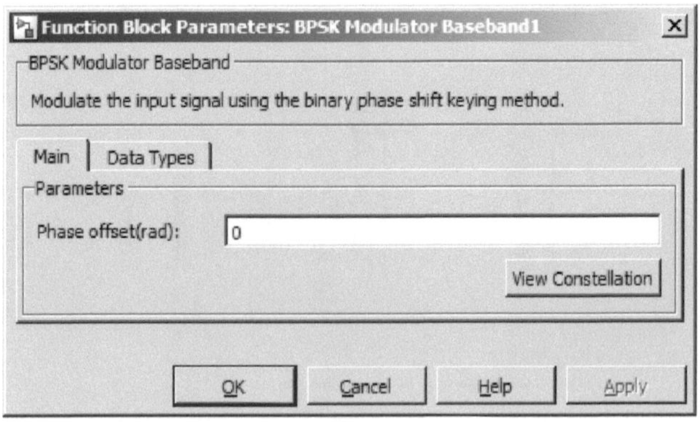

Figure 3.3 BPSK Modulator Input Parameters.

The BPSK demodulator menu is shown in Figure 3.6 where hard decisions are made at the demodulator output and the phase angle is again selected to be zero. The data type button shown in Figure 3.7 indicates that the output is also double precision, which is inherited from the preceding block.

Table 3.1 shows the first six values produced by each block based on a 5 s simulation time. The values are displayed in the work space by selecting

CONSTRUCTION OF A SIMULINK MODEL IN SIMPLE STEPS

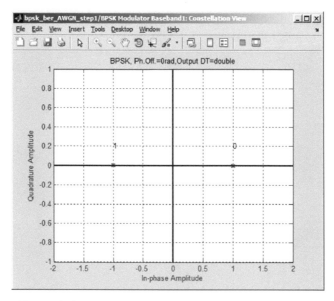

Figure 3.4 BPSK Constellation with Zero Phase Angle.

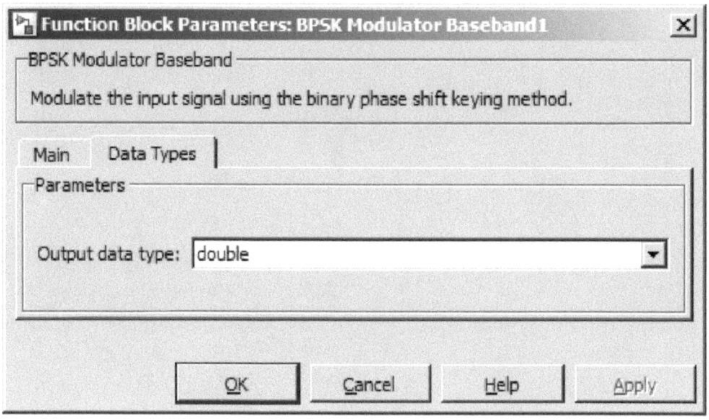

Figure 3.5 BPSK Output Data Type.

the tab with the corresponding label in the simulation. Note that with no additional noise the BPSK input and BPSK demodulator output produce the same sequence.

Figure 3.8 displays a modified simulation model where an AWGN block is introduced between the modulator and demodulator and a scope is included to compare input and output data sequences. The routing symbols labeled S are connectors for the data, used to avoid cluttering the model, with an extra line.

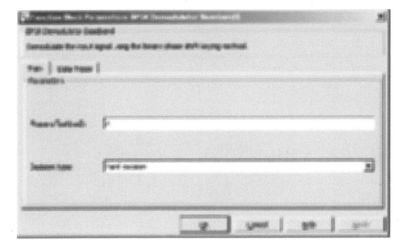

Figure 3.6 BPSK Demodulator Input parameters.

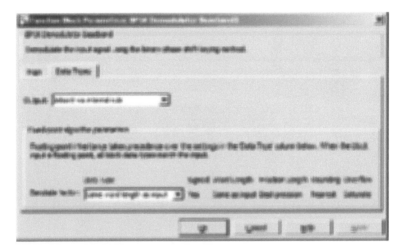

Figure 3.7 BPSK Demodulator Output Data Type.

The parameter selection for the AWGN block is shown in Figure 3.9 where the initial seed is 67, the number of bits/symbol is 1, the symbol period is 1 s and the signal power is 1 W. The E_b/N_o is selected to be 100 dB to demonstrate that the AWGN block introduces no errors with this large E_b/N_o value. The simulation time is extended to 100 s allowing the input and output sequences to be observed in the scope display as seen in Figure 3.10. The sequences in

TABLE 3.1 BPSK Modulator and Demodulator Outputs

Time (s)	0	1	2	3	4	5
bpsk_in	1	0	0	0	1	1
bpsk_mod_out	−1.0000 + 0.0000i	1.0000 + 0.0000i	1.0000 + 0.0000i	1.0000 + 0.0000i	−1.0000 + 0.0000i	−1.0000 + 0.0000i
bpsk_demod_out	1	0	0	0	1	1

50 DIGITAL COMMUNICATIONS BER PERFORMANCE IN AWGN (BPSK & QPSK)

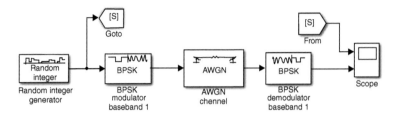

Figure 3.8 Simulink Model for BPSK (Step 2).

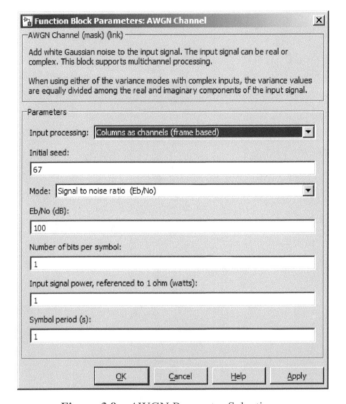

Figure 3.9 AWGN Parameter Selections.

Figure 3.10 are random and agree symbol by symbol due to the high E_b/N_o and are represented as double precision, real values that are either zero or one.

The E_b/N_o value has been changed to $-10\,\text{dB}$ in Figure 3.11 where it is observed that numerous errors occur.

CONSTRUCTION OF A SIMULINK MODEL IN SIMPLE STEPS 51

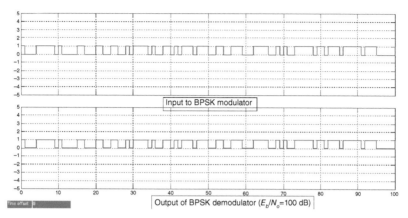

Figure 3.10 BPSK Modulator Input and BPSK Demodulator Output ($E_b/N_o = 100$ dB).

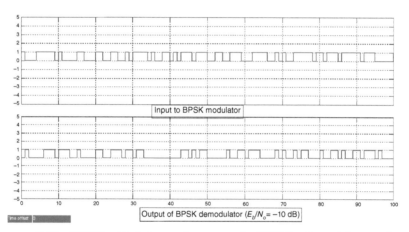

Figure 3.11 BPSK Modulator Input and BPSK Demodulator Output ($E_b/N_o = -10$ dB).

In Figure 3.12, the real and imaginary parts of the input and output of the AWGN block are displayed in the additional scopes. Figures 3.13 displays the real and imaginary parts of the BPSK modulator output using scope 2 where it is observed that the BPSK modulator produces real antipodal signals that are ±1 corresponding to the random source outputs 0 and 1. Using scope 3, Figure 3.14 displays the real and imaginary parts of the of the AWGN block output for $E_b/N_o = 4$ dB. Figure 3.15 displays the input to the BPSK modulator and the output from the BPSK demodulator on Scope 1. It is observed that

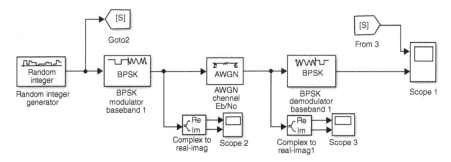

Figure 3.12 Simulink Model for BPSK (Step 3).

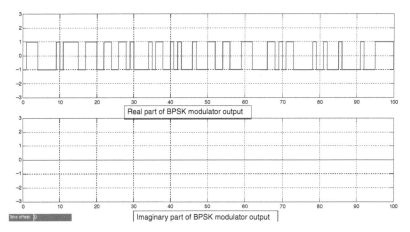

Figure 3.13 Real and Imaginary Parts of the BPSK modulator Output.

with $E_b/N_o = 4\,\text{dB}$ few demodulation errors are made so that the input and output sequences are the same; if the simulation time were extended, errors would be apparent.

The next step in this simulation, shown in Figure 3.16, is to include the error rate calculation block and its associated display. The parameter selections in the error rate calculation block are shown in Figure 3.17. The receive delay and computation delay are both set to zero for this example.[2] The display block indicates the estimated error rate, the number of errors, and the total number of symbols used in the simulation The upper number in display block displays a BER = 0.0127 for the specified $E_b/N_o = 4\,\text{dB}$;

---

[2]The receive delay is set when there is a known lag in the receive data. The computation delay allows transient behavior in the received data to be excluded from the BER estimate.

CONSTRUCTION OF A SIMULINK MODEL IN SIMPLE STEPS

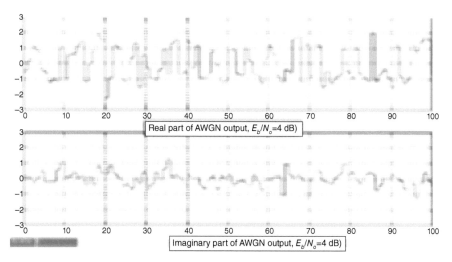

Figure 3.14 Real and Imaginary Parts of the AWGN Output ($E_b/N_o = 4$ dB).

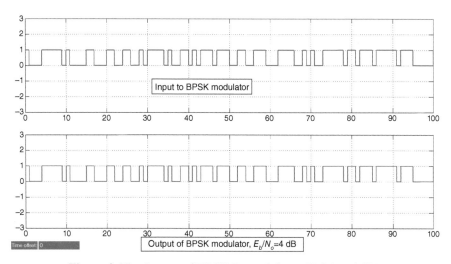

Figure 3.15 Output of BPSK Demodulator ($E_b/N_o = 4$ dB).

1270 errors are produced in the simulation using 100,000 symbols sent and received.

The final step in the construction of this model is to display the signal and port data types as shown in Figure 3.18. This selection is available in the Simulink model window menu under Display. An information block is also introduced to record principal parameters such as the 100,000 s simulation time, the 1 s symbol time, the 1 W signal power, and $E_b/N_o = 4$ dB.

54 DIGITAL COMMUNICATIONS BER PERFORMANCE IN AWGN (BPSK & QPSK)

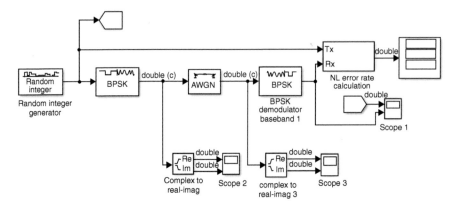

Figure 3.16 Simulink Model for BPSK (Step 4).

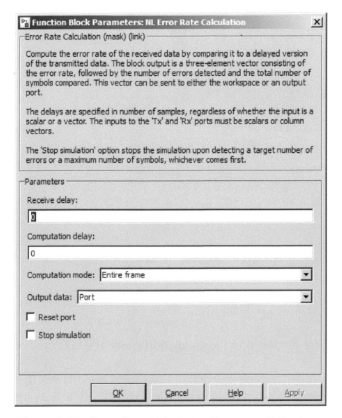

Figure 3.17 Error Rate Calculation Parameter Selections.

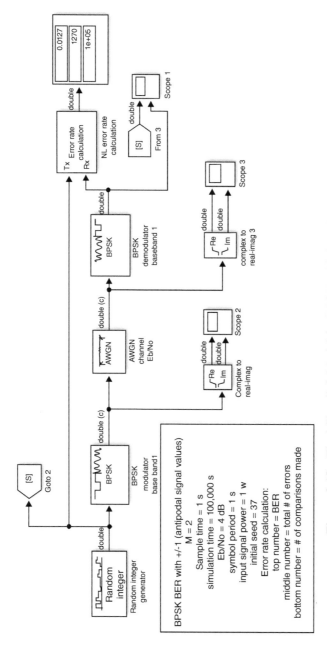

Figure 3.18 Final Simulink Model for BPSK with Signal and Port Data Types (Step 5).

3.3 COMPARISON OF SIMULATED AND THEORETICAL BER

Parameters used in the Simulink model for the BPSK BER simulation, shown in Figures 3.16 and 3.17, are specified as follows:

- BPSK antipodal signals = +1 and −1 ($M = 2$)
- Symbol period = 1 s
- Sample time = 1 s
- Run time = 100,000 s
- Random integer seed = 37
- Input signal power = 1 W
- AWGN random seed = 67
- Receive and computation delay = 0
- AWGN with γ_b = 4 dB =>
  - 1270 errors in 100,000 s
  - simulated BER = 0.0127

This model uses complex, sample-based signals where the sample time[3] is selected to be 1 s with one sample per symbol and antipodal BPSK symbols are generated. In general, the sample time may be changed to explore the vagaries of the channel, which may change at much smaller time intervals. Often an increase in the number of samples per symbol is desired as the case occurs when synchronization techniques are being investigated.

In this simulation, the AWGN block used E_b/N_o as the specified mode. Figure 3.19 indicates alternative mode choices are available such as symbol based E_s/N_o. For this binary case, $E_s/N_o = E_b/N_o = $ SNR where the symbol time equals the sample time and the time bandwidth product equals 1.

The theoretical BER of BPSK is well known and can be compared with the simulated results for specific energy contrast ratios denoted as $\gamma_b = E_b/N_o$. Appendix 3.A presents a short review of the theoretical BPSK BER performance. P_b, the probability of bit error for BPSK in AWGN, is given by the

---

[3] In the MathWorks help file sample time and sample rate are defined as follows: "A discrete-time signal is a sequence of values that correspond to particular instants in time. The time instants at which the signal is defined are the signal sample times, and the associated signal values are the signal samples. For a periodically sampled signal, the equal interval between any pair of consecutive sample times is the signal sample period, T_s. The sample rate, F_s, is the reciprocal of the sample period. It represents the number of samples in the signal per second."

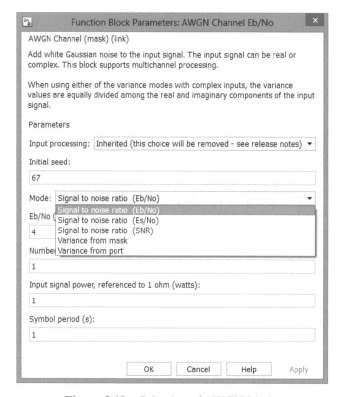

Figure 3.19 Selection of AWGN Mode.

relation

$$P_b = \frac{1}{2}\text{erfc}(\sqrt{\gamma_b})$$

where $\gamma_b = E_b/N_o$.

As an example with $\gamma_b = 4$ dB, $P_b = 0.0125$, a number that agrees reasonably with the simulated BER = 0.0127. Note that the accuracy of the simulated result improves with an increasing number of demodulated symbols.

A comparison of the theoretical and simulated BER for several values of γ_b is shown in Figure 3.20 for an simulation time of 100,000 s. The plot is most easily obtained by using the Mathworks bertool where a range of E_b/N_o values is entered and both theoretical and simulated results can be plotted. Table 3.2 lists simulated and theoretical BER values for specific values of γ_b. In the simulated case, a better estimate of the BER at low BER is obtained by using a larger number of transmitted symbols. Typically, the number of symbols should be selected to be an order of magnitude larger than the inverse of the BER, for example, 10^6 symbols should be sent for a BER = 10^{-5}.

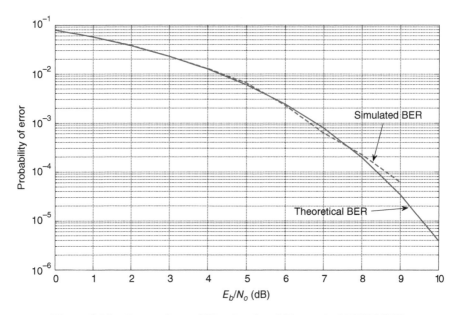

Figure 3.20 Comparison of Simulated and Theoretical BPSK BER.

TABLE 3.2 Theoretical and Simulated BER Using 100,000 Transmitted Symbols

γ_b	Theoretical P_b	Simulated P_b
0	0.0786	0.0777
1	0.0563	0.0566
2	0.0375	0.0381
3	0.0229	0.0229
4	0.0125	0.0127
5	5.95×10^{-3}	6.4×10^{-3}
6	2.39×10^{-3}	2.26×10^{-3}
7	7.73×10^{-4}	6.2×10^{-4}
8	1.91×10^{-4}	2.2×10^{-4}
9	3.36×10^{-5}	6.0×10^{-5}

3.4 ALTERNATE SIMULINK MODEL FOR BPSK

The AWGN block is very convenient to use when a BER curve is to be computed using the bertool and where E_b/N_o is a parameter. There are situations where it is not possible to simply introduce the AWGN block in the simulation and the noise must be added directly. Another example where the Gaussian noise block must be used is the case when the Gaussian noise is actually the

ALTERNATE SIMULINK MODEL FOR BPSK

source. However, it is important to understand that under the proper simulation conditions both the AWGN block and the Gaussian noise block used with an adder produce the same results. Note that the Gaussian noise block must have zero mean and a noise variance that corresponds to the desired value of E_b/N_o

The Mathworks documentation provides the relationship for the noise variance for the real and imaginary parts of complex Gaussian noise as

$$\text{Complex noise variance} = \frac{1}{2} \frac{\text{Signal power} \times \text{Symbol period}}{\text{sample time} \times 10^{0.1 \times E_s/N_o}}$$

where E_s/N_o is the ratio of the symbol energy to noise spectral density in dB. For example, in a binary case, if the signal power is 1 W, the symbol time and sample time are each 1 s, $E_b/N_o = E_s/N_o = 3$ dB results in a complex noise variance of 0.25 for both the real and imaginary parts.

Now an alternate model that uses Gaussian noise blocks instead of the AWGN block is shown in Figure 3.21. This model performs the same BPSK BER simulation as in the AWGN case where the simulation model parameters are specified as follows.

- BPSK antipodal signals $= +1$ and -1 ($M = 2$)
- Symbol period $= 1$ s
- Sample time $= 1$ s
- Run time $= 100,000$ s
- Random integer seed $= 37$
- Input signal power $= 1$ W
- Gauss noise random seeds $= 43$ and 37
- Mean of Gauss noise $= 0$
- Variance of Gauss noise $= 0.25$ for real and imaginary parts
- For $\gamma_b = 3$ dB $\Rightarrow$ theoretical $P_b = 0.0229$

This model uses a complex Gaussian noise generator in place of the AWGN block with the parameters indicated in Figure 3.22. Note that the seed numbers are different for the real and imaginary parts of the complex Gaussian noise generator to ensure statistical independence between the quadrature noise components. Running variance blocks are used to compute the signal power for several signals, which are then displayed in the model. From the displays, it can be observed that the noise power from display 2 is 0.4994 and the signal power from display 1 is 1.0 so that the approximate signal to noise ratio $\gamma_b = 3$ dB. From display 3, the total signal plus noise power is estimated

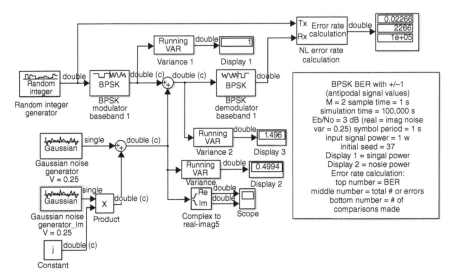

Figure 3.21 BPSK BER Simulation Using Complex Gaussian Noise.

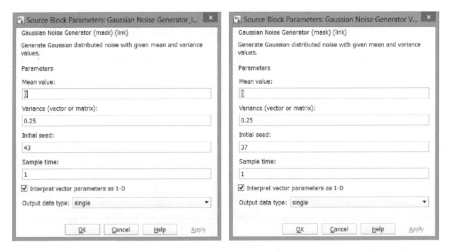

Figure 3.22 Gaussian Noise Block Parameters for Real and Imaginary Parts.

to be 1.496. The simulated BER is 0.02266, and is close to the theoretical BER = 0.0229 for $\gamma_b = 3\,\text{dB}$. Figure 3.23 illustrates the real and imaginary parts of the Gaussian noise generator for a variance equal to 0.25 used in both the real and imaginary Gaussian noise blocks.

The model shown in Figure 3.24 is simpler than that in Figure 3.21 and produces the same BER. It is not apparent from the model and the signal powers displayed that these models should produce the same result. The

ALTERNATE SIMULINK MODEL FOR BPSK

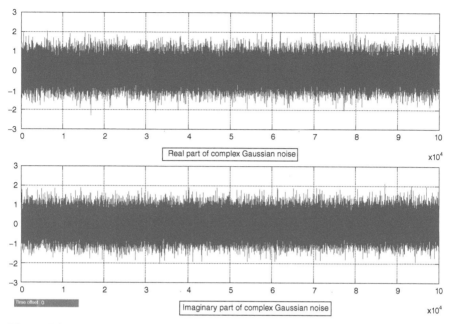

Figure 3.23 Real and Imaginary Parts of Gaussian Noise Generator (Var = 0.25 for both).

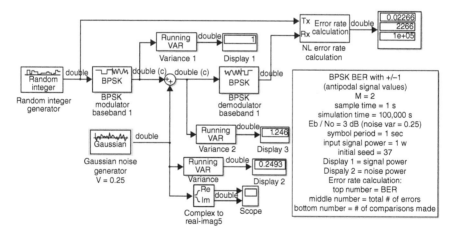

Figure 3.24 BPSK BER Simulation Using a Single Gaussian Noise Block.

underlying reason that the same BER is computed is due to the fact the BPSK demodulator operates only on the real part of the received signal. The imaginary part is discarded in the demodulator so that including the imaginary Gaussian noise block is unnecessary.

An alternate BPSK BER Simulink model that provides a direct comparison of use of the AWGN and Gaussian noise generator blocks is shown in Figures 3.25. For $\gamma_b = 3$ dB both methods of simulating the noise result in estimated BER values, that is, 0.02287 and 0.02266, each very close to theoretical BER = 0.0229. It should be noted that complex Gaussian noise is not needed since the BPSK demodulator takes the real part of the input signal to form its decision.

3.5 FRAME-BASED SIMULINK MODEL

There are instances when it is desirable to use frame-based rather than sample-based computation. A frame consists of a sequential sequence of samples from a single channel or multiple channels; the user must specify the frame size as an integer number of samples. In Simulink, the frame status is symbolized by a single line, →, for a sample-based signal and a double line, ⇒, for a frame-based signal.

Frame-based simulations execute faster and are often needed for matrix computations, where it is inconvenient to use sample-based computation. An example of a BPSK BER frame-based simulation is provided in Figure 3.26 with parameters specified as follows:

- BPSK antipodal signals = +1 and −1
- Symbol period = 1 s
- Sample time = 1 s
- Number of samples/frame = 10
- Run time = 100,000 s
- Random integer seed = 37
- Input signal power = 1 W
- AWGN random seed = 67
- AWGN with $\gamma_b = 3$ dB => theoretical $P_b = 0.0229$

Figure 3.27 displays the selection of 10 samples/frame while retaining a sample time equal to 1. The output power from the BPSK modulator = 1 and the output power from the AWGN block = 1.505 and are observed to be the same as the corresponding outputs in Figure 3.25. The simulated BER = 0.02287 is identical to that obtained in the sample-based model using the AWGN path in Figure 3.25.

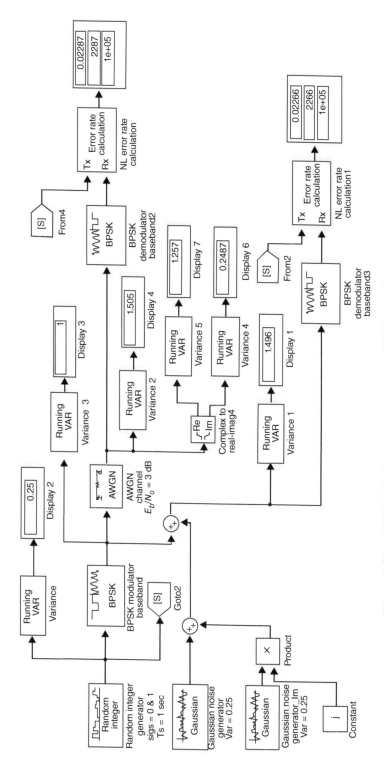

Figure 3.25 BPSK BER Simulation with both AWGN and Gaussian Noise Blocks.

63

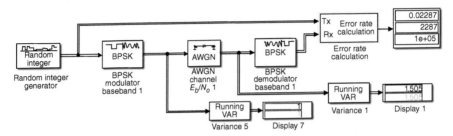

Figure 3.26 Frame-Based BPSK BER Simulation.

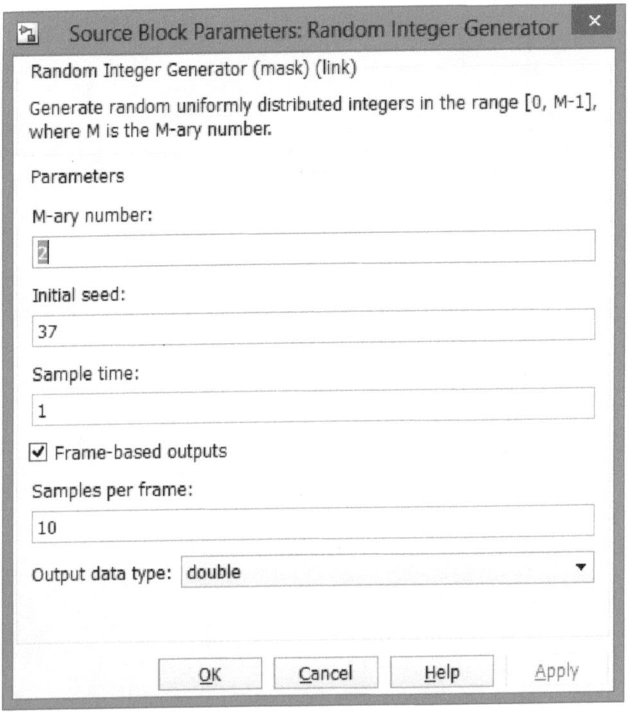

Figure 3.27 Source Block with 10 Samples per Frame.

3.6 QPSK SYMBOL ERROR RATE PERFORMANCE

Figure 3.28 presents a model for estimating QPSK BER. In this model, additional blocks, identified as scatter blocks, display the constellation of the QPSK demodulator and the received input to the QPSK demodulator. The random integer source specifies $M = 4$ with a sample time = 1 s. The parameter selections for the QPSK modulator are shown in Figure 3.29; here $M = 4$ and the phase angles are offset by $\pi/4$ using Gray coding. Figure 3.30 displays

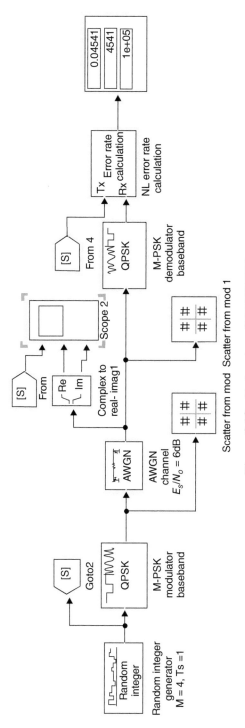

Figure 3.28 Estimation of QPSK BER.

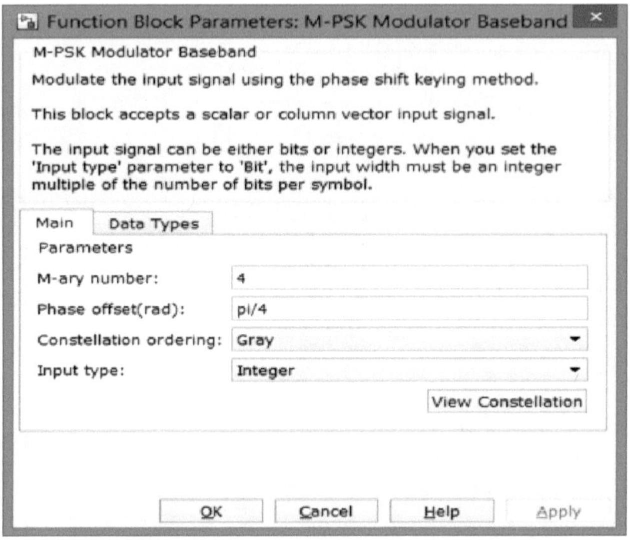

Figure 3.29 QPSK Modulator Parameters.

Figure 3.30 AWGN Parameter Selections for QPSK.

QPSK SYMBOL ERROR RATE PERFORMANCE

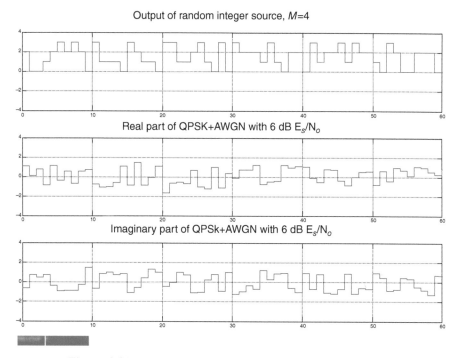

Figure 3.31 Scope 2 Display of Source and AWGN Outputs.

the AWGN parameter choices where it is seen that the symbol period is 1, the signal power is 1 W and the symbol signal to noise, $E_s/N_o = 6\,\text{dB}$.

From Scope 2, a segment of the source output and the real and imaginary parts of the AWGN output with $E_s/N_o = 6\,\text{dB}$ are shown in Figure 3.31.

The constellations for the QPSK modulator and the output of the AWGN block are shown in Figure 3.32. The QPSK phase angles are seen to be located at $+\pi/4, 3\pi/4, -3\pi/4, -\pi/4$. On the right side of Figure 3.32, multiple signal plus noise samples are displayed due to the simulation time of 100,000 s.

Theoretical QPSK symbol error rate performance denoted by P_s is known to be

$$P_s = erfc(\sqrt{\gamma_b})\left[1 - \frac{1}{4}erfc\left(\sqrt{\gamma_b}\right)\right]$$

where γ_b is the SNR per bit, E_b/N_o. Since QPSK symbols have two bits per symbol, the symbol SNR, $E_s/N_o = 2E_b/N_o$ or $\gamma_s = 2\gamma_b$. Then for $\gamma_b = 3\,\text{dB}$ or $\gamma_s = 6\,\text{dB}$, the theoretical QPSK symbol error rate in AWGN is 0.0455. The simulated symbol error rate for $\gamma_s = 6\,\text{dB}$ is 0.04541 based on the 100,000 s simulation time and agrees reasonably with the theoretical symbol error rate.

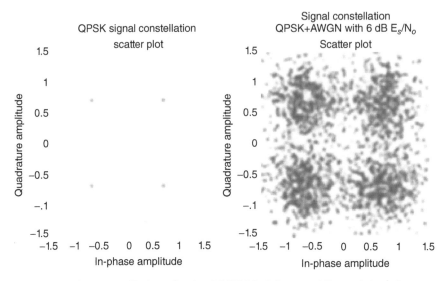

Figure 3.32 Constellations for the QPSK Modulator and Demodulator Input.

Figure 3.33 provides a model for determining QPSK symbol error rate performance using both AWGN and Gaussian noise blocks. In this model, each variance of the real and imaginary parts of the complex Gaussian noise is 0.125; with unity signal power the symbol SNR $\gamma_s = 6$ dB.

In summary, the symbol error rate for QPSK with Gray coding and $\gamma_s = 6$ dB is provided as follows where good agreement is observed between theory and simulation:

Theoretical with Gaussian noise $\gamma_s = 6$ dB ($\gamma_b = 3$ dB)
- $P_s = 0.0455$

Simulated with AWGN $\gamma_s = 6$ dB
- $P_s = 0.04541$

Simulated with Gauss, Real Part = Imaginary Part Var = 0.125
- $P_s = 0.04546$

3.7 BPSK FIXED POINT PERFORMANCE

The previous simulations had available theoretical results allowing theory and simulated performance to be compared directly. In these cases, the power of Simulink is not apparent. Often it is difficult or impossible to obtain theoretical results so that simulation becomes the clear choice for estimating

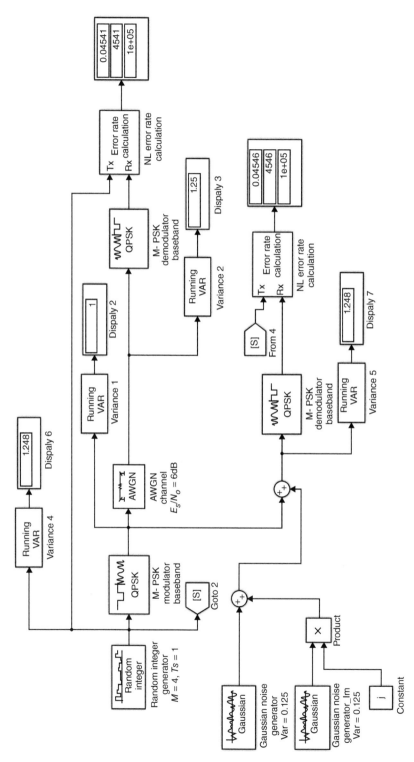

Figure 3.33 QPSK Symbol Error Rate Using AWGN and Gaussian Noise Blocks.

performance. An example will now be provided where fixed point arithmetic required for digital signal processing (DSP) implementations (such as in an Application-specific Integrated Circuit (ASIC) or an Field-programmable Gate Array (FPGA)) is utilized in the model.

Fixed point numbers are represented in binary by means of their word length denoted by ws, a binary point and a fraction of length, n. The fixed point structure, shown in the Figure 3.34 diagram, places the most significant bit (MSB) left most and the least significant bit (LSB) right most. The MSB is assumed to be a sign bit, s, that is assigned 1 for signed representation and 0 for unsigned. The discussion provided here follows the MATLAB designation where the fixed point number is expressed as fixdt(s,ws,n). MATLAB restricts the fixed length word size to be 128 bits or less. The selection of the word size and fraction length is based on the eventual implementation in an ASIC or FPGA device where the word size controls the range of the values to be represented and the fraction length controls the precision. More detail on fixed point representations is available in the MathWorks documentation under the category fixed point numbers.

Figure 3.35 compares a fixed point simulation (bottom of model) to a floating point simulation (top of model) for BPSK in AWGN with $\gamma_b = 3$ dB and an execution time of 100,000 s.

In the model shown in Figure 3.35, the convert block translates the floating point numbers to fixed point precision using fixdt (1,8,4) corresponding to a word length ws = 8, fraction length $n = 4$ and signed bit $s = 1$. The input parameters for the convert block are shown in Figure 3.36. Since the BPSK modulator would normally be part of the fixed point implementation, the same fixed point representation fixdt(1,8,4) is used. However, use of fixed point in the BPSK modulator does not affect the BER performance. In Simulink, the AWGN block requires a floating point input which explains the need for the double block that precedes the AWGN block. The results shown in Figure 3.35 indicate that with $\gamma_b = 3$ dB the fixed point error rate = 0.02332 and is only slightly poorer than floating point error rate = 0.02287. (Note that the six digit theoretical BER for $\gamma_b = 3$ dB is 0.022878.)

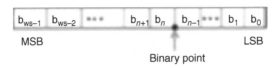

Figure 3.34 Fixed Point Number Representation.

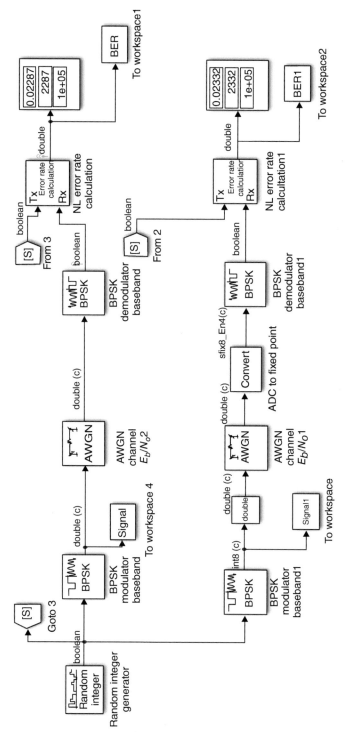

Figure 3.35 Fixed & Floating Point BPSK Simulation.

Figure 3.36 Convert Block Parameter Settings.

The choice of word length and fraction length is usually a compromise involving the numerical range of the numbers to be represented, the precision of these numbers and the quantization errors introduced to deliver the best representation of the digitized signals for the planned hardware implementation. Saturation may occur if the upper and lower limits of the numbers are exceeded with the consequence that inaccurate estimation or unpredictable errors occur.

An example of fixed point issues identified thus far is obtained by computing BER results for selected fixed point combinations using the model shown in Figure 3.35. Figure 3.37 plots the error rate for fixed point cases with ws = 8 and fraction lengths of 1, 2, and 4. Use of a fraction length greater than 4 causes saturation in this model. Note that with γ_b = 10 dB the BER for the case fixdt (1,8,4) results in slightly degraded performance compared to theoretical. This result is attributed to an insufficient execution time for adequately

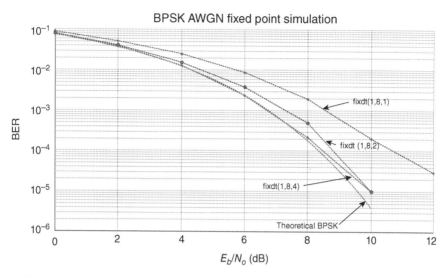

Figure 3.37 Comparison of Fixed Point BER with Theoretical BPSK BER.

estimating the BER performance rather than from errors introduced by quantization noise.

3.8 SUMMARY DISCUSSION

Theoretical BPSK BER performance in AWGN is well known. Simulation results using AWGN blocks and Gaussian noise generator blocks yield identical performance with properly chosen parameters. Use of a large sample size produces excellent agreement between theoretical and simulated BER performance. The AWGN block or a Gaussian noise block may be selected for either sample-based or frame-based simulation. The BPSK and QPSK simulations that were developed do not fully demonstrate the power of Simulink when the theoretical performance is known. A fixed point example is included where theoretical performance is difficult or impossible to obtain.

APPENDIX 3.A THEORETICAL BER PERFORMANCE OF BPSK IN AWGN

The BER performance for BPSK is obtained using the following low pass equivalent signals:

- received signal, $r(t)$

- transmitted signal $u_i(t)$ with bit energy E_b
- additive white Gaussian noise $z(t)$ with zero mean and power spectral density N_o

With these definitions the received signal in AWGN is given by

$$r(t) = \alpha e^{-j\phi} u_i(t) + z(t), \ 0 \le t \le T, \ i = 1, 2$$

where α is a constant, T is the symbol duration, φ carrier phase and one of two transmitted signals is transmitted. In the case of antipodal signals $u_1(t) = -u_2(t)$. The bit energy is expressed as $E_b = \frac{1}{2} \int_0^T |u_i(t)|^2 dt$. Assuming coherent detection where the phase is known at the receiver, the decision variables are represented by

$$U_1 = 2\alpha E_b + \mathrm{Re}\left\{ e^{j\varphi} \int_0^T z(t) u_1^*(t) dt \right\}$$

$$U_2 = 2\alpha E_b \rho_r + \mathrm{Re}\left\{ e^{j\varphi} \int_0^T z(t) u_2^*(t) dt \right\}$$

where ρ_r, the real part of the signal cross correlation, is

$$\rho_r = \frac{1}{2E_b} \mathrm{Re}\left\{ \int_0^T u_1(t) u_2^*(t) dt \right\}$$

The probability of bit error, P_b, is then found to be

$$P_b = \frac{1}{2}\mathrm{erfc}\left(\sqrt{\frac{\gamma_b}{2}(1 - \rho_r)} \right)$$

where the energy contrast ratio $\gamma_b = \alpha^2 \frac{E_b}{N_0}$. For antipodal signals $\rho_r = -1$.
For notational convenience in this chapter, the constant $\alpha = 1$.
In summary, the BER for BPSK in AWGN with antipodal signals is

$$P_b = \frac{1}{2}\mathrm{erfc}(\sqrt{\gamma_b})$$

where $\gamma_b = \frac{E_b}{N_0}$.

PROBLEMS

3.1 Reduce the symbol time to 0.1 s in the Figure 3.18 Simulink model using $E_b/N_o = 3$ dB with an execution time $=100,000$ s.

  a. Does the BER remain the same as the case with 1 s symbol duration?

  b. Why does it take longer to execute with 0.1 s symbols?

3.2 Using the symbol time as 0.1 s in the Figure 3.18 Simulink model increase $E_b/N_o = 10$ dB with an execution time $= 100,000$ s. Does the BER remain the same as the case with 1 s symbol duration using $E_b/N_o = 10$ dB?

3.3 Using the symbol time as 0.1 s in the Figure 3.18 Simulink model with $E_b/N_o = 10$ dB, select frame-based simulation with 10 samples per symbol

  a. Is the BER the same as the case with sample-based simulation?

  b. Does the simulation take the same time to complete?

3.4 Figure P.3.1 presents a simulation where the channel introduces a fixed phase offset. In this simulation, the phase offset is set to 30° and the BER is sent to the work space for a range of values between 0 and 10 dB. The BER results are plotted in Figure P.3.2 where the bertool is used to compute the theoretical and simulated BER cases. The results show the degradation from a fixed channel phase offset compared to the case where no offset exists.

Repeat this simulation where the 30° phase offset is introduced in the demodulator and the channel phase offset is set to zero. Does this produce the same results?

3.5 In Figure P.3.3, a sine wave is added to filtered Gaussian noise with the following parameters:

- Gauss Noise block: zero mean, var $= 0.25$, sample time Ts $= 0.01$ s, frame based with 1000 samples/frame
- Low pass filter block: Filter specs: FIR, minimum, single rate; freq specs: normalized 0 to 1, fpass $= 0.45$, fstop $= 0.55$; magnitude specs: dB units, Astop $= 1$, Apass $= 60$, equiripple
- Sine wave block: amplitude $= 0$, frequency $= 10$ Hz, phase $= 0$, sample time $T_s = 0.01$ s, frame based with 1000 samples/frame

  a. Execute the simulation using 1000 s and observe the spectrum analyzer output. Observe that the spectrum analyzer output extends

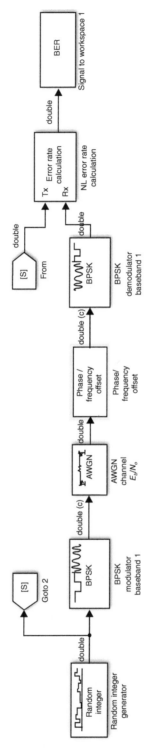

Figure P.3.1 Channel Introduces Fixed Phase Offset.

PROBLEMS

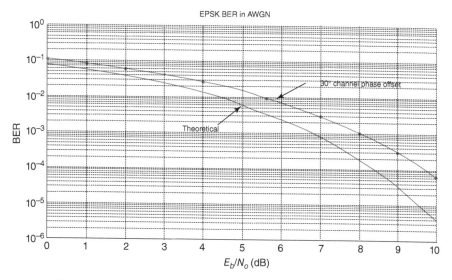

Figure P.3.2 BPSK BER Performance with Fixed Phase Offset.

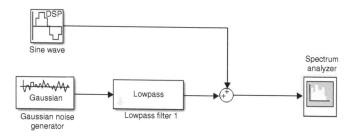

Figure P.3.3 Filtered Sine plus Gaussian Noise Simulink Model.

from -50 to 50 Hz and explain what sets these limits. What are the frequency limits for the lowpass filter?

b. Add a running variance block at the output of the Gaussian noise block, the output of the sine wave block, the output of the low pass filter, and the output of the adder. What power levels are obtained?

c. Make the following changes in the low pass filter: fpass = 0.2, fstop = 0.3. What power levels are obtained? Explain what the differences are due to compared with part a.

3.6 Show that the BER at 10 dB in the Figure 3.35 model can be improved by a longer execution time.

3.7 Show that using fixdt(1,8,5) in the Figure 3.35 model causes saturation.

4

DIGITAL COMMUNICATIONS BER PERFORMANCE IN AWGN (MPSK & QAM)

4.1 MPSK AND QAM ERROR RATE PERFORMANCE IN AWGN

This chapter continues the development of Simulink models for coherent modulations including multi-phase PSK (MPSK) and quadrature amplitude modulation (QAM). Topics presented here are listed as follows:

- MPSK BER Performance (floating point)
- MPSK BER Performance (fixed point)
- QAM BER Performance
  - Average power
  - Peak power
  - Nonlinear amplifiers

4.2 MPSK SIMULINK MODEL

In MPSK, it is assumed that k information bits are assigned to $M = 2^k$ equal-energy signals. As a specific example, Figure 4.1 displays the Simulink model

---

Modeling of Digital Communication Systems Using SIMULINK®, First Edition.
Arthur A. Giordano and Allen H. Levesque.
© 2015 John Wiley & Sons, Inc. Published 2015 by John Wiley & Sons, Inc.
Companion Website: www.wiley.com/go/simulink

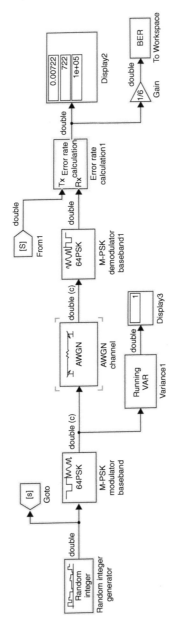

Figure 4.1 MPSK Simulink Model with $M = 64$.

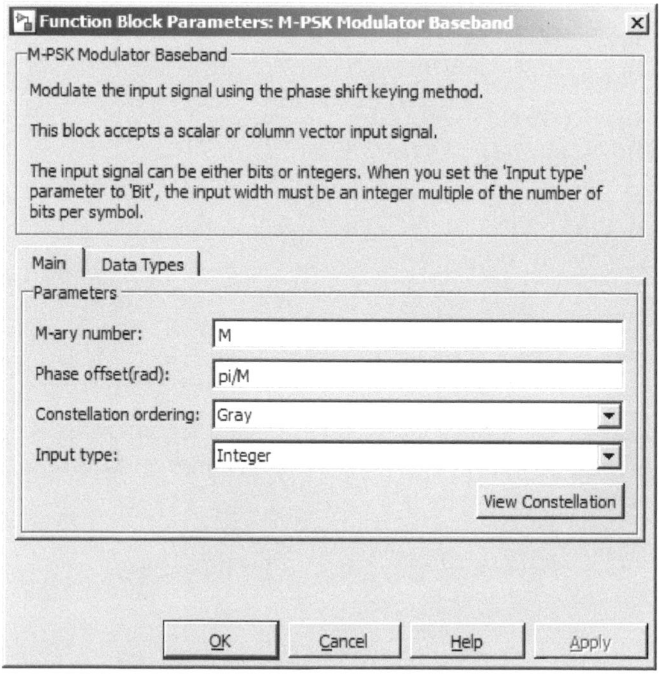

MPSK SIMULINK MODEL

Figure 4.2 MPSK Input Parameters.

for determining the symbol error rate (SER) and BER performance for MPSK modulation with $k = 6$ and $M = 64$.

The parameters in the MPSK modulator block are displayed in Figure 4.2. Note that in this simulation the parameters M and k must be entered in the MATLAB command window prior to executing the model. Gray coding is stipulated as well. The AWGN block parameters are displayed in Figure 4.3.

In the AWGN block, E_s/N_o(dB) is selected as γ_s (dB) $= \gamma_b$ (dB) $+ 10\log(k)$. For $k = 6$ and $\gamma_b = 24$, γ_s(dB) $= 24 + 7.815 = 31.7815$ dB. Parameters and performance results used in the Simulink model for the MPSK simulation, shown in Figure 4.1, are specified as follows:

- $M = 64$; $k = 6$
- Symbols $= +\pi/64 \ldots -\pi/64$; Gray coding
- Symbol period $= 1$ s
- Sample time $= 1$ s

- Run time = 100,000 s
- Random integer seed = 37
- Input signal power = 1 W
- AWGN random seed = 67
- Receive and computation delay = 0 s
- AWGN with $\gamma_s = 31.78$, ($\gamma_b = 24\,\text{dB}$) = >
  - 722 errors in 100,000 s
  - simulated SER = 0.00722
  - simulated BER = 0.00120

For comparison with the simulated BER, the theoretical BER for MPSK in AWGN is presented here. With $M = 2^k$ the theoretical SER denoted by P_s, for MPSK is approximated by

$$P_s = \text{erfc}\left(\sqrt{k\gamma_b}\sin\left(\frac{\pi}{M}\right)\right) \text{ for } \gamma_b \gg 1$$

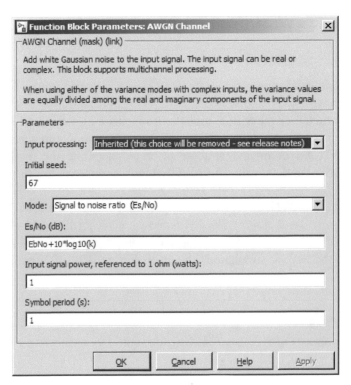

Figure 4.3 AWGN Parameters for MPSK Simulation.

FIXED POINT BER FOR MPSK

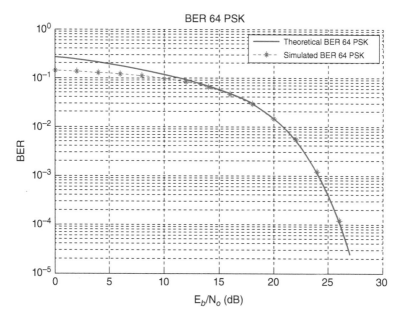

Figure 4.4 BER for 64 PSK in AWGN.

The BER is well approximated for high SNR by $P_b = \frac{P_s}{k}$. Note that the symbol $E_s/N_o = \gamma_s$ is related to the $E_b/N_o = \gamma_b$ according to $\gamma_s = k\gamma_b$

The bertool is now used to produce the BER plot shown in Figure 4.4. Note that the approximation BER $P_b = \frac{P_s}{6}$ becomes very accurate as the signal to noise ratio increases.

4.3 BER FOR OTHER ALPHABET SIZES

The BER results for other alphabet sizes are easily obtained from the Simulink model shown in Figure 4.1 by specifying other values of M and k. As an example, the constellation for 8-PSK is shown in Figure 4.5.

The BER results for $M = 4, 8, 16, 32$, and 64 are now obtained using the bertool with M and k entered in the MATLAB command window as needed. Figure 4.6 plots the theoretical and simulated BER versus Eb/No for MPSK in these cases. Note that $M = 2$ and $M = 4$ exhibit the same BER.

4.4 FIXED POINT BER FOR MPSK

BER results for MPSK using fixed point arithmetic in AWGN are provided here. Since Simulink supports fixed point alphabets with only $M = 2, 4$, or 8,

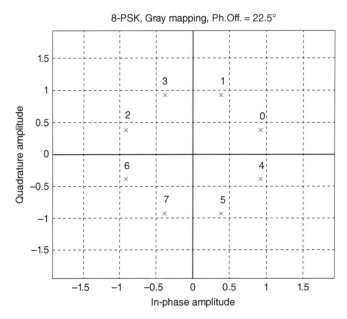

Figure 4.5 Constellation for Octal PSK.

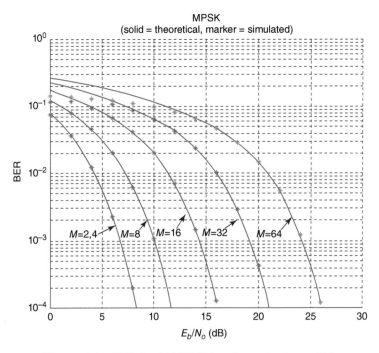

Figure 4.6 BER for MPSK Using Several values of M.

the example provided here assumes $M = 8$. The Simulink model for $M = 8$ is shown in Figure 4.7.

Using the bertool, BER results are obtained for a fixed-point word length $= 8$ and fraction lengths of 2, 3, and 4. The BER results shown in Figure 4.8 are obtained using a simulation time $= 100,000$ s. It is observed that use of a fraction length $= 3$ provides the closest results to theoretical BER.

4.5 QAM SIMULINK MODEL

Simulink models for computing the QAM BER performance in AWGN are developed here. An important choice in simulating the QAM BER performance is whether the performance is obtained for peak or average power. Peak power is often an important consideration when the power must be constrained to stay within the power amplifier (PA) limits to avoid saturation. In the next example, BER is computed using average power; subsequent examples will demonstrate the need for peak power computations.

The constellation with Gray coding for $M = 64$ with $k = 6$ is shown in Figure 4.9.

QAM model parameters and resulting BER performance are specified as follows:

- $M = 64; k = 6$
- Gray coding
- Symbol period $= 1$ s
- Sample time $= 1$ s
- Run time $= 1,000,000$ s
- Random integer seed $= 37$
- Average signal power $= 1$ W
- AWGN random seed $= 67$
- Receive and computation delay $= 0$ s
- AWGN with $\gamma_b = 15$ dB or $\gamma_s = 22.78$ dB $=>$
  - 781 errors in 1,000,000 s
  - simulated BER $= 0.00078$

The QAM modulation function block parameters are shown in Figure 4.10. The normalization can be set to average power, peak power, or minimum distance between symbols.

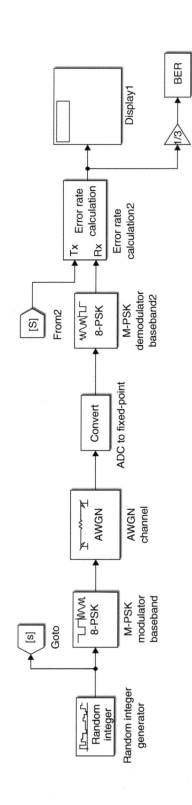

Figure 4.7 Model for Simulation of Octal PSK BER with Selected Fixed-Point Fraction Lengths.

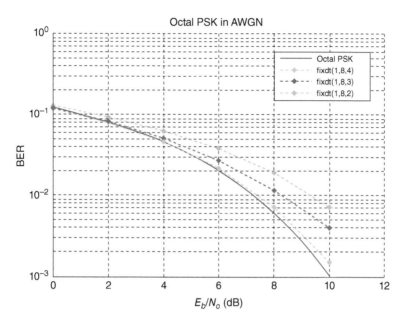

Figure 4.8 BER for Octal PSK in AWGN with Word Length 8.

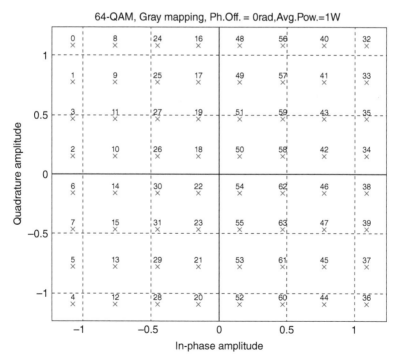

Figure 4.9 QAM Constellation for $M = 64$.

Figure 4.10 QAM Modulation Parameters with Average Power.

The Simulink model for computing the QAM BER with $M = 64$ and $k = 6$ is shown in Figure 4.11. In the simulation $\gamma_s = 22.78$ dB corresponding to $\gamma_b = 15$ dB (Note that $\gamma_s = \gamma_b + 10\log(6)$). The simulated BER result based on a simulation time $= 1{,}000{,}000$ s is then 0.00078.

For comparison with the simulation results, the theoretical QAM BER is presented here. If P_s denotes the probability of symbol error, then P_b, the probability of bit error, with Gray coding is approximately $P_b \approx \frac{P_s}{k}$. For rectangular signal constellations with $k = \log_2 M$ and k even, the symbol error probability is

$$P_s = 2\left(1 - \frac{1}{\sqrt{M}}\right)\mathrm{erfc}\left(\sqrt{\frac{3k\gamma_b}{2(M-1)}}\right)$$

$$\times \left\{1 - \frac{1}{2}\left(1 - \frac{1}{\sqrt{M}}\right)\mathrm{erfc}\left(\sqrt{\frac{3k\gamma_b}{2(M-1)}}\right)\right\}$$

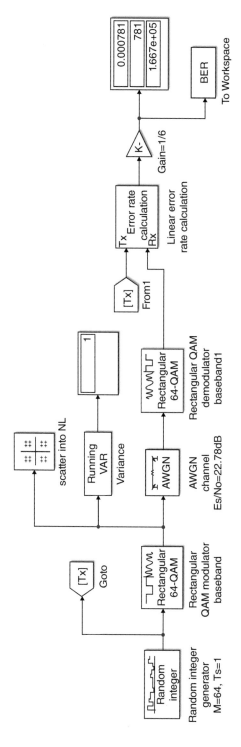

Figure 4.11 Simulink Model to Compute QAM BER.

For k odd P_s is upper bounded by

$$P_s \leq 2\mathrm{erfc}\left(\sqrt{\frac{3k\gamma_b}{2(M-1)}}\right)$$

With $M = 64$, $k = 6$, and $\gamma_b = 15$ dB, $P_b = 0.000771$. It is now seen that the simulated and theoretical results are in good agreement.

4.6 QAM BER FOR OTHER ALPHABET SIZES USING AVERAGE POWER

The model shown in Figure 4.11 can be modified to generate results for other alphabet sizes. The simulation results for QAM with $M = 4$, 16, and 64 assuming average power and a simulation time $= 10^6$ s are shown in Figure 4.12. Good agreement is observed between the theoretical and simulation results.

4.7 QAM BER USING PEAK POWER

To compute the QAM BER with peak power, for both the modulator and demodulator, the peak power must be selected as shown for the modulator

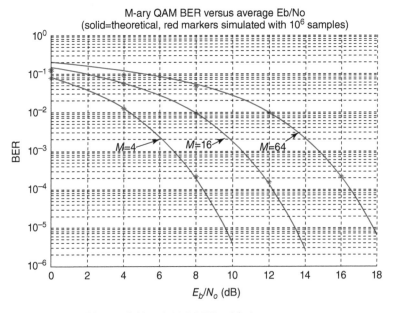

Figure 4.12 QAM BER with Average γ_b.

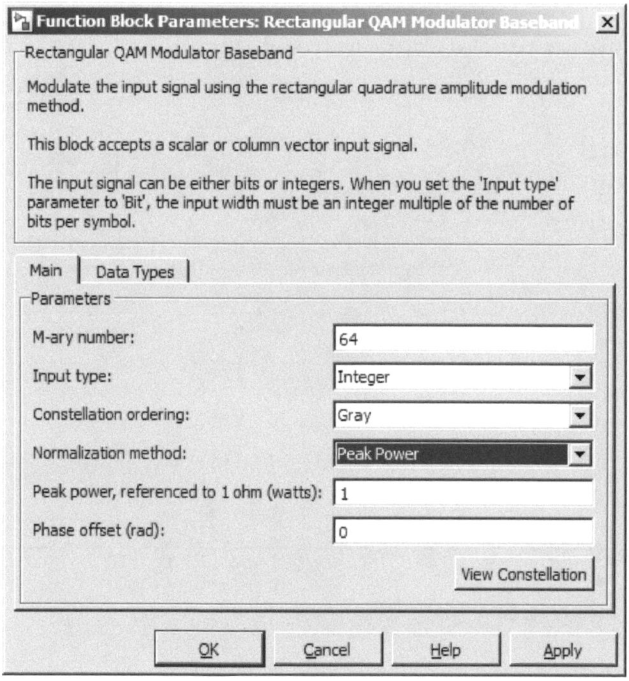

Figure 4.13 QAM Modulator Input Parameters with Peak Power Set to 1 W.

in Figure 4.13. A comparison of the simulated QAM BER for $M = 64$ using both average and peak power is shown in Figure 4.14 for a simulation time of 10^6 s; the solid line shows the $M = 64$ theoretical QAM BER results for average power. With a rectangular constellation and $M = 64$ the degradation between average and peak power is computed to be 3.68 dB. The degradation is most easily observed by examining Figure 4.14 at a BER = 2×10^{-4}.

4.8 POWER AMPLIFIER CONSTRAINT USING PEAK POWER SELECTION WITH QAM

In satellite applications, the PA is implemented as a traveling-wave tube (TWT) or a solid state power amplifier (SSPA) and is ordinarily operated in its linear region by ensuring that the peak power is well below the nonlinear PA saturation point. As a result, a penalty in BER is incurred relative to average power operation when the results are compared with those in the previous section.

In general, power amplifiers exhibit memoryless, nonlinear behavior. Several models for representing the characteristics of power amplifiers are

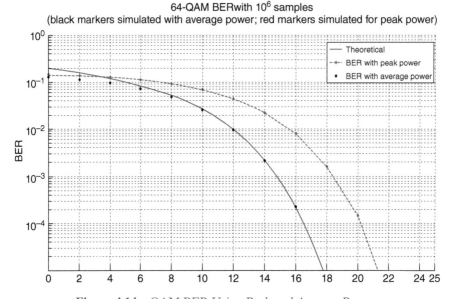

Figure 4.14 QAM BER Using Peak and Average Power.

presented in the MathWorks documentation. One of the power amplifier models, attributable to Saleh[1], will be studied here to estimate the impact on BER performance. This example highlights the utility of Simulink for a practical case where no theoretical results are available.

The implementation to be studied involves the insertion of a nonlinear device after the QAM modulator causing the modulated waveform to experience AM/AM and AM/PM conversion. In the Saleh model, an analytic representation of these conversions utilizes four parameters, that is, α_a and β_a for the AM/AM conversion and α_p and β_p for the AM/PM conversion. The AM/AM and AM/PM equations are expressed in Saleh's model in terms of the magnitude of the input voltage u as

$$F_{AM/AM} = \frac{\alpha_a u}{1 + \beta_a u^2}$$

$$F_{AM/PM} = \frac{\alpha_p u^2}{1 + \beta_p u^2}$$

---

[1] Saleh, A.A.M., "Frequency-independent and frequency-dependent nonlinear models of TWT amplifiers," IEEE Transactions on Communications, vol. COM-29, pp. 1715–1720, November 1981.

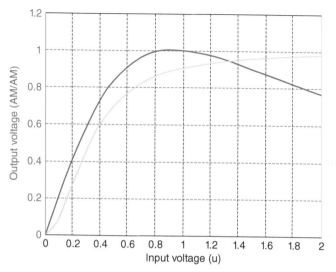

Figure 4.15 Saleh Nonlinear Model for AM/AM and AM/PM Conversion.

respectively, where $F_{AM/AM}$ is the output voltage versus the input voltage for the AM/AM conversion and $F_{AM/PM}$ is the output phase versus the input voltage for the AM/PM conversion

Figure 4.15, extracted from the MathWorks documentation, displays graphs of these equations where specific values of the parameters suggested by Saleh are $\alpha_a = 2.1587$, $\beta_a = 1.1517$, $\alpha_p = 4.0033$, and $\beta_p = 9.1040$. Saleh notes that an rms error of 0.010 occurs for the AM/AM parameters and 0.469 (in degrees whereas the AM/PM formula is in radians) for the AM/PM parameters. These equations and values of the parameters have been found to fit measured TWT data.

Figure 4.16 depicts a BER simulation using 64-QAM with and without a nonlinear PA based on the Saleh model. Peak power is selected in the QAM modulator and demodulators, $\gamma_b = 15$ dB and the simulation time = 100,000 s. The BER without the nonlinear device is 0.014 and with the nonlinear device is 0.165 thus indicating a significant penalty imposed by the nonlinear device.

Figure 4.17 illustrates the scatter plots at the input and ouput of the Saleh nonlinearity. The signal constellation at the output shows scattering and warping of the rectangular 64-QAM constellation resulting in a poor BER.

When a QAM modulator output is the input to a nonlinear device, the input voltage is backed off to force the modulated waveform to remain within the linear portion of the device. To avoid sacrificing available transmit power, a predistortion device is inserted prior to the nonlinearity to compensate for the

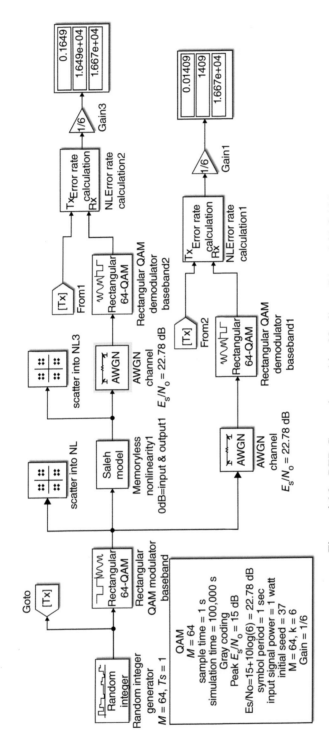

Figure 4.16 BER Comparison with a Nonlinear PA and 64-QAM.

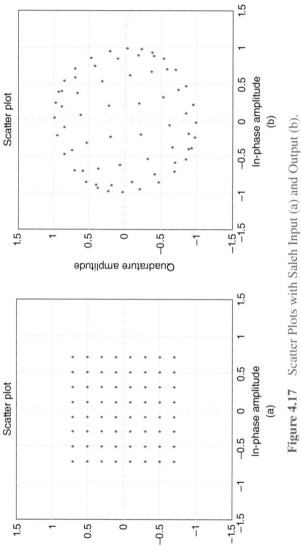

Figure 4.17 Scatter Plots with Saleh Input (a) and Output (b).

distortion as long as the saturation point is not exceeded. Since the Saleh model is an ideal monotonic function up to the saturation point, an ideal predistortion device can be implemented that exactly compensates for the distortion by computing an inverse function of the nonlinear characteristic.

Incorporation of a predistortion device has the added benefit of demonstrating the use of a MATLAB S function defined by the user. Specifically, the S function presented here accomplishes the exact compensation and is a perfect linearizer. The S function, labeled nlinvd.m, is implemented in MATLAB. The m-file listing is given as follows:

```
%Inverse Saleh AM/AM & AM/PM
function inl = nlinvd(u);
%u2 = u^2;
%fam = 2.1587*u/(1+1.1517*u2);
%fpm = 4.0033*u2/(1+9.1040*u2);
%aa = .8*2.1587;
%ba = .8*1.1517;
aa = 2.1587;
ba = 1.1517;
ap = 4.0033;
bp = 9.1040;
magam = abs(u);
xu = aa/magam;
nu = (xu-(xu^2-4*ba)^.5)/(2*ba);
nu2 = nu^2;
nang = ap*nu2/(1+bp*nu2);
nuang = angle(u)-nang;
inl = nu*exp(j*nuang);
```

Figure 4.18 introduces the predistortion block between the QAM modulator output and the Saleh nonlinearity. Peak power is again selected in the QAM modulator and demodulators where the input voltage magnitude $|u| = 1$ is used to avoid saturation, $\gamma_b = 15$ dB and the simulation time $= 100{,}000$ s. The BER without the nonlinear device and with the included predistortion block is 0.014 and with the nonlinear device is 0.165. It is now evident that the predistortion device compensates exactly for the Saleh nonlinearity. The scatter plots, shown in Figure 4.19, at the input to the predistortion block and the input and output of the Saleh nonlinearity also demonstrate that the input signal constellation and Saleh output constellation are the same. Note that the actual implementation of the predistortion device will degrade BER performance.

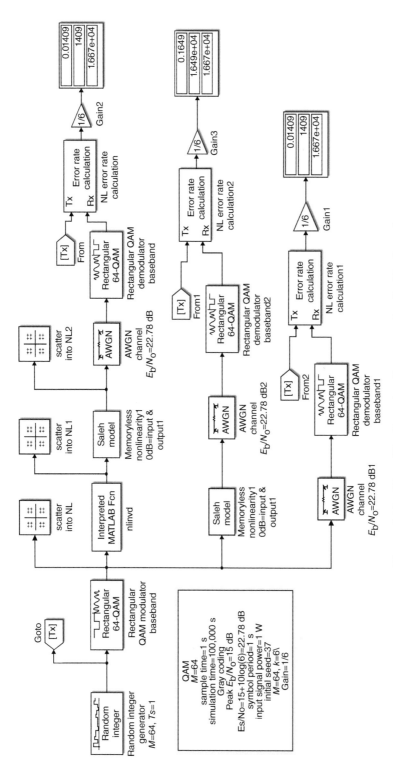

Figure 4.18 Use of Predistortion Compensation for Nonlinearity.

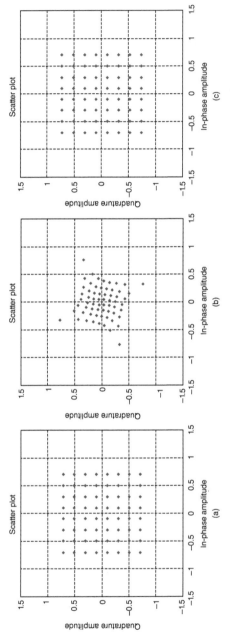

Figure 4.19 Scatter Plots: Predistortion Input (a), Saleh Input (b), and Saleh Output (c).

4.9 SUMMARY DISCUSSION

This chapter has developed theoretical and simulated BER results for MPSK and QAM. The use of the bertool was applied extensively in the BER computations. QAM signaling required a study of both peak and average power to address issues with saturation in nonlinear PAs. The power of Simulink was revealed in the example where a nonlinear device was introduced in the simulation to address the case where no theoretical results are available. The incorporation of a user-defined S function, developed in MATLAB, was also illustrated to allow the user to modify the simulation when no library block is available.

PROBLEMS

4.1 Modify the parameters in the Simulink model shown in Figure 4.1 to reproduce the curves in Figure 4.6.

4.2 Using Simulink show that $M = 2$ and $M = 4$ produce about the same BER.

4.3 Using the QAM Simulink model, determine the simulated BER for $M = 256$ and compare the result to the theoretical performance.

4.4 For $M = 16$ compute the simulated QAM BER for γ_b in the range from 10 to 15 dB assuming both peak and average power. What is the theoretical peak power degradation? What is the approximate degradation obtained in the simulation?

4.5 In the Figure 4.16 model change the method in the Saleh block to cubic polynomial with the third order intercept point = 30 dBm, AM/PM conversion = 0° per dB, lower input power = 10 dBm, and upper input power = inf. Assume $\gamma_b = 15$ dB and the simulation time = 100,000 s.

a. What is the simulated BER in this case?

b. Show the scatter plot at the nonlinearity output.

c. Change the third order intercept point to 35 dBm and determine the simulated BER and the scatter plot at the nonlinearity output.

5

DIGITAL COMMUNICATIONS BER PERFORMANCE IN AWGN (FSK AND MSK)

5.1 FSK AND MSK ERROR RATE PERFORMANCE IN AWGN

This chapter addresses constant envelope modulations including binary frequency-shift keying (BFSK), multiple frequency-shift keying (MFSK), offset QPSK (OQPSK) and bandwidth-efficient modulations, minimum-shift keying (MSK) and Gaussian minimum shift keying (GMSK). Specific results are presented for

- BFSK BER performance
- MFSK BER performance
- MSK, OQPSK, and GMSK BER performance
- MSK and GMSK power spectra

5.2 BFSK SIMULINK MODEL

A simulation to estimate the BER for BFSK with noncoherent detection in AWGN is shown in Figure 5.1.

Modeling of Digital Communication Systems Using SIMULINK®, First Edition.
Arthur A. Giordano and Allen H. Levesque.
© 2015 John Wiley & Sons, Inc. Published 2015 by John Wiley & Sons, Inc.
Companion Website: www.wiley.com/go/simulink

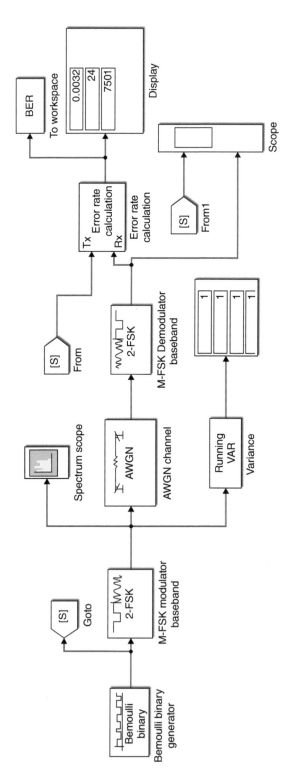

Figure 5.1 Model for Simulation of BFSK BER in AWGN with Noncoherent Detection.

BFSK SIMULINK MODEL

Specific input parameters are as follows:

- BFSK noncoherent detection
- Symbol period = 0.2 s
- Sample time = symbol period = 0.2 s
- Run time = 1500 s
- Bernoulli binary seed = 61
- Frame based with 1 sample/frame
- Frequency spacing = 200 Hz (±100 Hz tones)
- FSK mod/demod = 2000 samples/symbol (f_s = 10 kHz)
- Input signal power = 1 W
- RBW = 976.56 mHz (~1 Hz)
- Receive and computation delay = 0
- AWGN with γ_b = 10 dB =>
  - simulated BER = 0.0032

The BFSK boundaries for each symbol are selected to have continuous phase. Since there are 2000 samples/symbol with a 0.2 s symbol duration, the sampling rate is 10 kHz and is well above the Nyquist rate for the 200 Hz tone separation. In the simulation, the spectrum scope computes the FFT with a rectangular window, 100 spectral averages with no overlap, a 1280 FFT size and a 976.56 mHz resolution bandwidth (RBW) as shown in Figure 5.2. (These settings are obtained by selecting spectrum settings under the View tab in the spectrum scope.)

Figure 5.3 shows the expanded spectral result displayed in Figure 5.2.

Figure 5.4 displays the complex BFSK modulator signal output for one symbol using a time scope from the DSP System toolbox (not shown in Figure 5.1). Figure 5.5 extends the BFSK modulator output to display more than two symbol periods to enable the reader to see where the BFSK frequencies switch tones.

Using a utility block (not shown in Figure 5.1) to convert the complex BFSK modulator output to real and imaginary signals, Figures 5.6 and 5.7 display the real and imaginary parts of the BFSK modulator output, respectively, for two symbol periods.

As shown in Figure 5.7, the switch in BFSK tones is observed at the 0.2 s symbol boundary

Figure 5.8 displays the input of the BFSK modulator (top trace) and the output of the BFSK demodulator (bottom trace). With γ_b =10 dB no errors are observed over this interval.

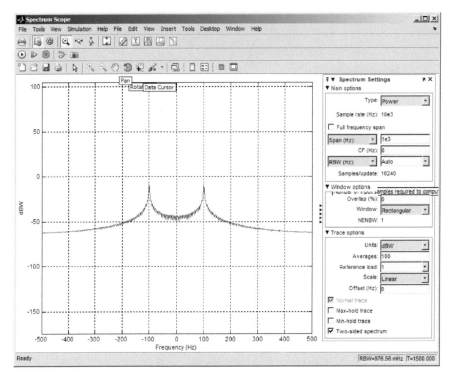

Figure 5.2 Spectrum Scope Display.

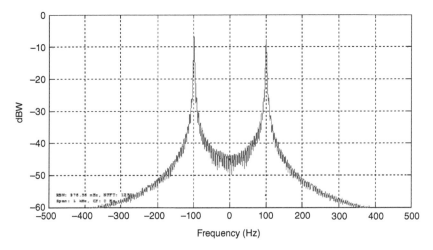

Figure 5.3 BFSK Modulator Spectrum with Tones at ± 100 Hz.

BFSK SIMULINK MODEL

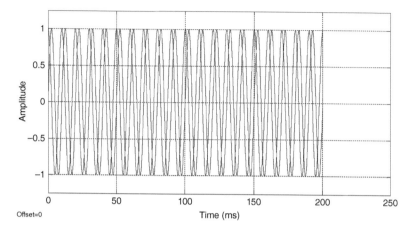

Figure 5.4 BFSK Modulator Output for 1 Symbol Period.

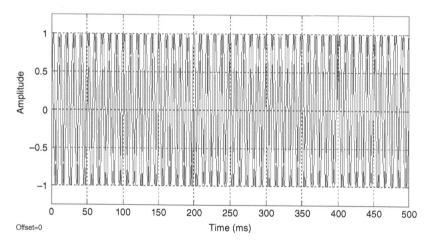

Figure 5.5 BFSK Modulator Output for More than 2 Symbol Periods.

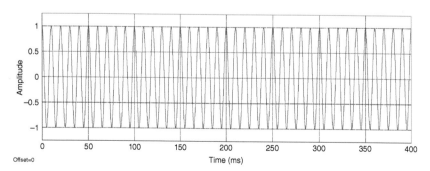

Figure 5.6 Real Part of BFSK Modulator Output.

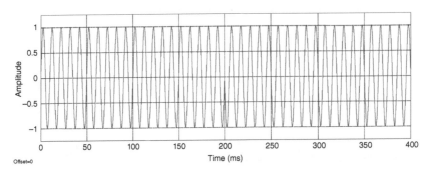

Figure 5.7 Imaginary Part of BFSK Modulator Output.

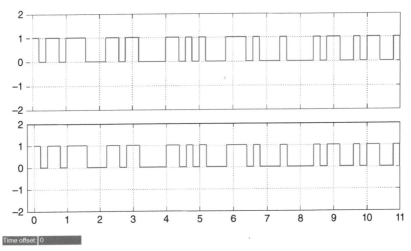

Figure 5.8 BFSK Modulator Input (Top Trace) and BFSK Demodulator Output (Bottom Trace).

The theoretical BER for BFSK in AWGN with noncoherent detection is given by

$$P_b = \frac{1}{2}e^{-\gamma_b/2}$$

Figure 5.9 is obtained by using the Simulink model in Figure 5.1 in conjunction with the bertool. Figure 5.9 then compares the simulated BER with the theoretical BER for BFSK in AWGN with noncoherent detection. Good agreement can be observed between the theoretical and simulated results.

MFSK SIMULINK MODEL

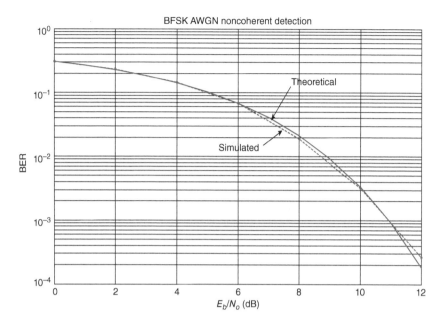

Figure 5.9 Simulated and Theoretical BFSK in AWGN.

5.3 MFSK SIMULINK MODEL

In this section, results are developed for BER performance in AWGN assuming noncoherent detection of M-FSK with $M = 2^k$. A Simulink model for 4-FSK is shown in Figure 5.10 where $\gamma_b = 6$ dB. Note that $\gamma_s = \gamma_b + 10\log(k) = 6 + 10\log(2) = 9$ dB. Specific input parameters are as follows:

- 4-FSK noncoherent detection
- Continuous phase tone boundaries
- Symbol period = 0.2 s
- Sample time = symbol period = 0.2 s
- Run time = 1500 s
- Random Integer seed = 37
- Frame based with 1 sample/frame
- Frequency spacing = 200 Hz (±100, ±300 Hz tones)
- FSK mod/demod = 2000 samples/symbol ($f_s = 10$ kHz)

- Input signal power = 1 W
- RBW = 5 Hz
- Gray coding
- Receive and computation delay = 0 s
- AWGN with $\gamma_b = 6$ dB, $\gamma_s = 9$ dB =>
  - simulated BER = 0.014

Figure 5.11 displays the spectrum of the 4-FSK modulator output using the Spectrum Scope. The sampling rate, based on 2000 samples/symbol and a 0.2-s symbol duration, is 10 kHz and remains above the Nyquist rate. In the simulation, the spectrum scope computes the FFT with a Hann window, 100 spectral averages with no overlap, a 501 FFT size and a 5 Hz RBW.

Figure 5.12 compares the input of the 4-FSK modulator (top trace) with the output of the 4-FSK demodulator (bottom trace). With $\gamma_b = 6$ dB an error occurs between 3 and 4 s.

Figure 5.13 shows theoretical and simulated BER performance for MFSK in AWGN with noncoherent detection and $M = 2, 4, 8, 16,$ and 32. Theoretical performance is given by

$$P_b = \frac{2^{k-1}}{2^k - 1} \sum_{n=1}^{M-1} (-1)^{n+1} \binom{M-1}{n} \frac{1}{n+1} e^{-nk\gamma_b/(n+1)}$$

Note that MFSK BER is obtained from the symbol error rate P_M with $M = 2^k$ according to $P_b = \frac{2^{k-1}}{2^k-1} P_M$. As an example with $k = 2$ and $M = 4$ the BER for 4-FSK is

$$P_b = e^{-\gamma_b} - \frac{2}{3} e^{-\frac{4}{3}\gamma_b} + \frac{1}{6} e^{-\frac{3}{2}\gamma_b}$$

For 4-FSK with $k = 2$ and $M = 4$ and $\gamma_b = 10$ dB, $P_b = 4.44 \times 10^{-5}$.

5.4 MSK SIMULINK MODEL

Minimum shift keying (MSK) is specific case of continuous phase frequency shift keying (CPFSK) where the peak frequency deviation is one half the bit rate. When this modulation is viewed as a variant of binary CPFSK, the tones are spaced at $+1/4T$ and $-1/4T$ from the carrier where T is the symbol duration and the frequency separation is $1/2T$. Alternatively MSK can be described as a form of offset quadrature phase shift keying (OQPSK) where the rectangular pulses are replaced by half sinusoidal pulses and the quadrature

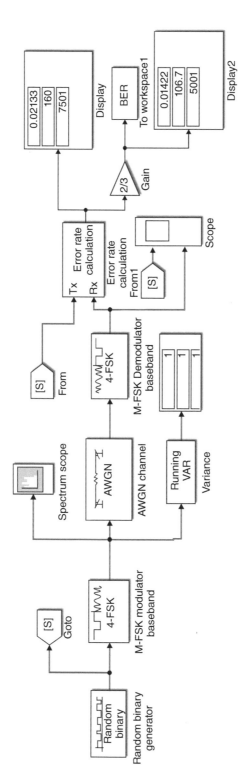

Figure 5.10 Model for Estimation of BER with 4-FSK in AWGN.

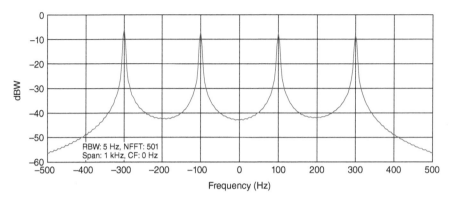

Figure 5.11 Spectrum of 4-FSK Modulator.

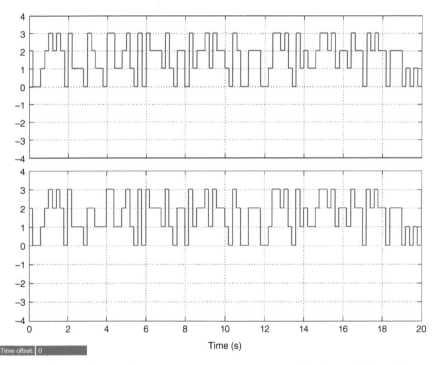

Figure 5.12 4-FSK Modulator Input (Top Trace) and 4-FSK Demodulator Output (Bottom Trace).

pulse is delayed by a half symbol from the in-phase pulse. Figure 5.14 illustrates the half symbol stagger between the in-phase and quadrature rectangular pulses in OQPSK. Figure 5.15 depicts MSK's half symbol staggered in-phase and quadrature sinusoidal pulses.

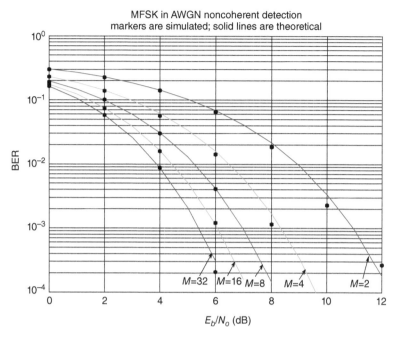

Figure 5.13 BER Performance for MFSK in AWGN.

A variation of MSK is Gaussian minimum shift keying (GMSK) where a premodulation filter having a Gaussian-shaped impulse response is inserted before the MSK modulator.

The Gaussian impulse response $g(t)$ is given by

$$g(t) = \frac{\sqrt{\pi}}{\alpha} \exp\left(-\frac{\pi^2}{\alpha^2}t^2\right)$$

where $\alpha = \frac{2\sqrt{\ln 2}}{B}$ and B is the 3 dB bandwidth. The filter transfer function, $G(f)$, is

$$G(f) = \exp(-\alpha^2 f^2)$$

The filter is then represented by its time bandwidth product BT where T is the symbol duration. As an example, $BT = 0.3$ for GSM cellular applications. The probability of bit error for MSK and GMSK (see Rappaport[1]) is expressed as

$$P_b = \frac{1}{2}\mathrm{erfc}(\sqrt{\alpha \gamma_b})$$

[1] Rappaport, T.A. Wireless Communications Prentice Hall, 1996 p. 264

112 DIGITAL COMMUNICATIONS BER PERFORMANCE IN AWGN (FSK AND MSK)

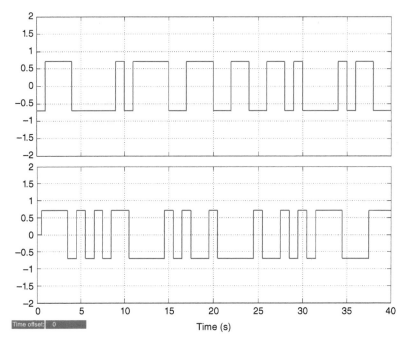

Figure 5.14 OQPSK In-Phase (Top Trace) and Quadrature (Bottom Trace) Pulses.

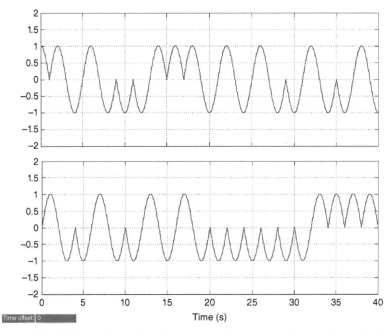

Figure 5.15 MSK In-Phase (Top Trace) and Quadrature (Bottom Trace) Pulses.

where $\alpha = 0.85$ for MSK with $BT = \infty$ and $\alpha = 0.68$ for GMSK with $BT = 0.25$. Note that the difference in γ_b between GMSK with $BT = 0.25$ and MSK is $10 \log(0.85/0.68) \approx 1$ dB.

A Simulink model that is used to estimate the BER for MSK, GMSK, and OQPSK in AWGN is displayed in Figure 5.16.

Figure 5.17 illustrates the BER for theoretical OQPSK and MSK BER and simulated BER results for OQPSK, MSK, and GMSK. The three Simulink models employ frame-based simulation with 16 samples/frame and an input signal power of 1 W. MSK and GMSK both use 1 sample/symbol and a traceback depth of 16. $BT = 0.25$ for GMSK.

Good agreement is observed between the theoretical and simulated BER for OQPSK and for MSK. The small loss in GMSK BER performance between GMSK and MSK is evident at low BER values.

5.5 MSK POWER SPECTRUM

Significant characteristics of MSK include the following features:

- Constant envelope (important for Class C nonlinear amplifiers)
- Spectrally efficient suppression of out-of-band interference
- Good BER performance

For the aforementioned reasons, MSK is a commonly implemented modulation. GMSK exhibits nearly the same BER as MSK but also offers improved spectral occupancy. Using the Simulink model in Figure 5.18, the power spectrum of MSK and GMSK is computed and displayed in Figure 5.19. The computations are frame based with 512 samples/frame, 4 samples/symbol for both MSK and GMSK, a time-bandwidth product $= 0.25$ for GMSK, a frequency span of 4 Hz and a 1 s symbol duration. The spectrum analyzer uses 100 averages with a Hann window and no overlap. Figure 5.19 shows that MSK exhibits higher sidelobes and a wider main lobe compared with GMSK. The constants in the simulation are selected to force the gain at zero frequency to be approximately 0 dB.

The theoretical power spectral density spectrum of MSK, denoted by $H(f)$, is expressed as[2]

$$H(f) = \frac{16PT}{\pi^2} \left(\frac{\cos(2\pi fT)}{1 - 16f^2T^2} \right)^2$$

---

[2] Sklar, B., Digital Communication, Prentice Hall, 2001 2nd ed, p. 562.

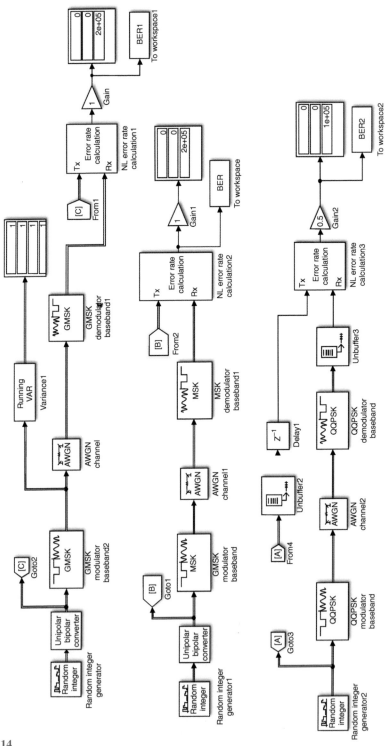

Figure 5.16 Simulink Model for BER Estimation of MSK, GMSK, and OQPSK in AWGN.

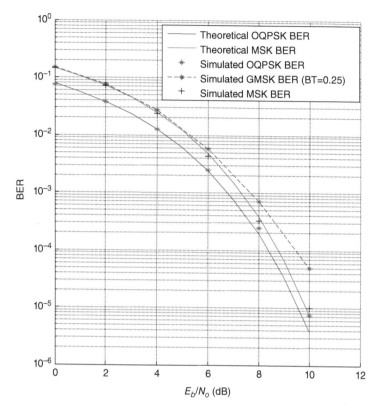

Figure 5.17 BER Performance for OQPSK, MSK, and GMSK in AWGN.

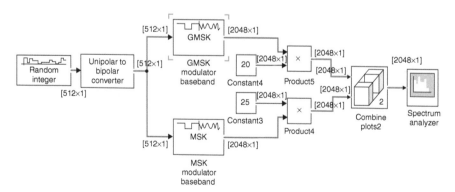

Figure 5.18 Simulink Model to Compute Power Spectrum for MSK and GMSK.

where P is the average power in the modulated waveform and T is the symbol time. With, $T = 1$ it can be seen that $H(f)$ has nulls at 0.75, 1.25, ... which are also evident in Figure 5.19.

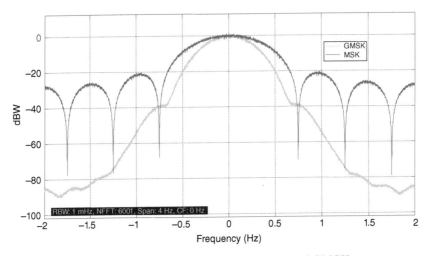

Figure 5.19 Power Spectrum for MSK and GMSK.

5.6 SUMMARY DISCUSSION

This chapter has continued the development of theoretical and simulated BER results, focusing here on BFSK, MFSK, MSK, and GMSK. The spectrum analyzer was used extensively and shown to offer a wide selection of spectral estimation techniques and parameters. The spectral efficiency of MSK and GMSK was illustrated and the spectrum of BFSK and 4-FSK were both displayed. GMSK and MSK were shown to offer comparable BER performance.

PROBLEMS

5.1 Replace auto in the spectrum scope with RBW = 5 Hz and compare the spectrum with Figure 5.3.

5.2 The power out of the modulator is 1 W. A sine wave has a power of 0.5 W. Explain why the modulator output power is 1 W.

5.3 Why does the real part not indicate a change in frequency from one symbol to the next whereas the imaginary part shows the change in frequency?

5.4 Modify the Simulink model in Figure 5.10 and reproduce the plot for $M = 32$.

5.5 Using the model in Figure 5.16, obtain a BER plot in the range $\gamma_b = 0$ to 10 dB for MSK and compare the simulated result with the theoretical MSK BER.

5.6 In the Simulink model in Figure 5.15, let $BT = 10$, change the GMSK scale to 25 and compute the power spectrum. How does this result compare with the MSK power spectrum?

6

DIGITAL COMMUNICATIONS BER PERFORMANCE IN AWGN (BPSK IN FADING)

6.1 BPSK IN RAYLEIGH AND RICIAN FADING

This chapter presents several topics in Simulink based on BPSK modulation in fading channels. Specifically these topics include the following:

- BPSK BER performance in Rayleigh fading
- BPSK BER performance in Rician fading
- BPSK BER performance in Rician fading with Multipath
- Rician channel fading characteristics

6.2 BPSK BER PERFORMANCE IN RAYLEIGH FADING

Prior chapters have introduced Simulink models at the outset to acquaint the reader with important techniques for Simulink model development. Here it is useful to review the theoretical BER performance for Rayleigh fading channels in order to establish a framework for Simulink model construction. Appendix 6.A presents a short review of BPSK BER performance in Rayleigh

---

Modeling of Digital Communication Systems Using SIMULINK®, First Edition.
Arthur A. Giordano and Allen H. Levesque.
© 2015 John Wiley & Sons, Inc. Published 2015 by John Wiley & Sons, Inc.
Companion Website: www.wiley.com/go/simulink

fading. The error probability for BPSK in Rayleigh fading is given by

$$P_b = \frac{1}{2}\left[1 - \sqrt{\frac{\overline{\gamma_b}}{1 + \overline{\gamma_b}}}\right]$$

where $\overline{\gamma_b}$ is the average SNR/bit. The time-varying nature of the channel is also characterized by its power spectral density $S(f)$. A specific model referred to as the Jakes model is given by

$$S(f) = \frac{1}{\pi f_m} \frac{1}{\sqrt{1 - (f/f_m)^2}} \quad |f| \leq f_m, \ 0 \ \text{otherwise}$$

where f_m is the maximum Doppler frequency.[1] In a mobile channel where a vehicle moves at a speed of v meters/second and the transmitted carrier frequency is f_0, then $f_m = v f_0/c$ where c is the velocity of light (3×10^8 m/s).

A Simulink model used to estimate BPSK BER performance in Rayleigh fading is shown in Figure 6.1.

A summary of the model parameters is specified as follows:

Model Parameters for BPSK in Rayleigh Fading

- Antipodal signals = +1 and −1, 1 bit/symbol
- Sample time = symbol time = 1 s
- Simulation time = 1,000,000 s
- Random integer seed = 22
- Jakes model with Doppler shift = 0.01 Hz
- Input signal power = 1 W
- Average SNR = 10 dB =>
  - Simulated BER = 0.0231
  - Theoretical BER = 0.0233

Simulation parameters for the Rayleigh fading block are displayed in Figure 6.2. Multipath parameters are not specified in this simulation since the Jakes model is a "flat" or frequency-nonselective model.

The AWGN block specifies the 1 W input signal power and the 1 s symbol time. The estimated BER is 0.023 with an average SNR $\overline{\gamma_b}$ = 10 dB. Figure 6.3 displays the real and imaginary parts of the Rayleigh fading channel gain.

---

[1] Proakis, J.G., and M. Salehi, Digital Communications, 5th ed, pp. 838–839.

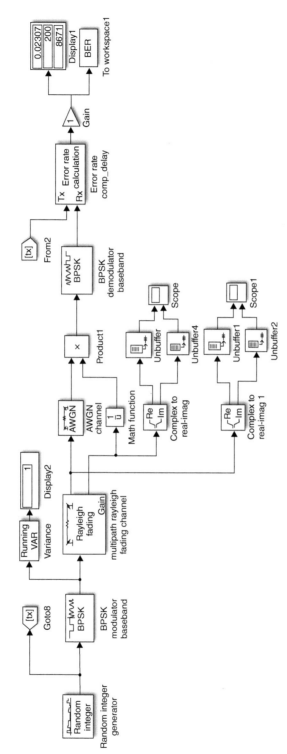

Figure 6.1 Simulink Model for Estimation of BPSK BER in Rayleigh Fading.

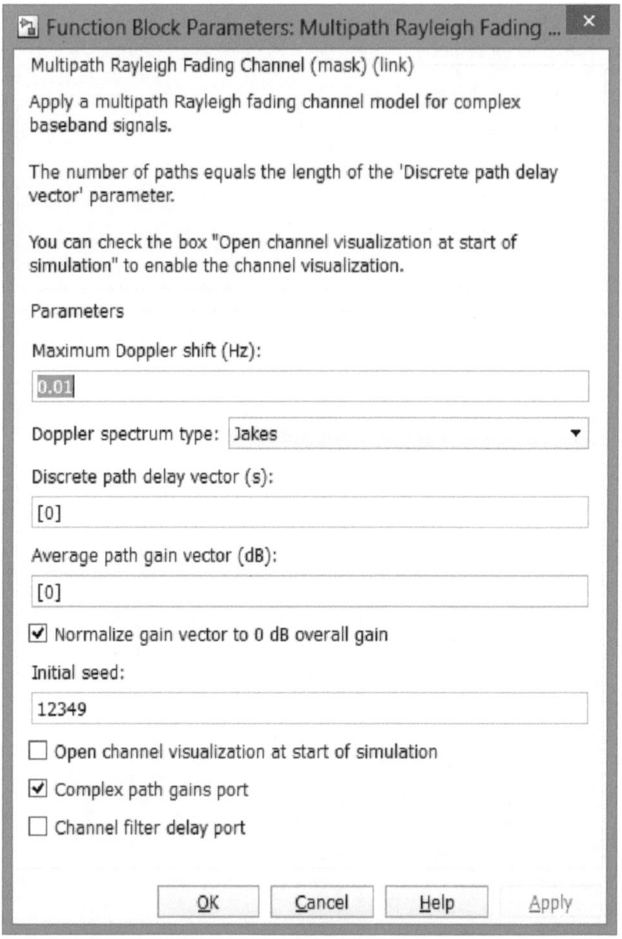

Figure 6.2 Rayleigh Fading Parameter Selection.

Figure 6.4 displays the real and imaginary parts of the Rayleigh fading channel output where the time variability of the channel is clearly evident as a result of the selection of the Jakes model with an 0.01 Hz maximum Doppler shift.

A notable block incorporated in this simulation is the math function $1/u$. This function is required to track the channel time-variability where the receiver implementation ordinarily incorporates an automatic gain control (AGC).

Figure 6.5 depicts results for BPSK BER in Rayleigh fading compared with BPSK BER in AWGN along with simulated results for BPSK in Rayleigh fading. The results are obtained using the bertool where the channel type is

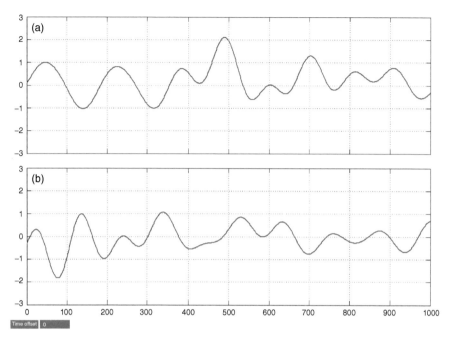

Figure 6.3 Real (a) and Imaginary (b) Parts of the Rayleigh Channel Gain.

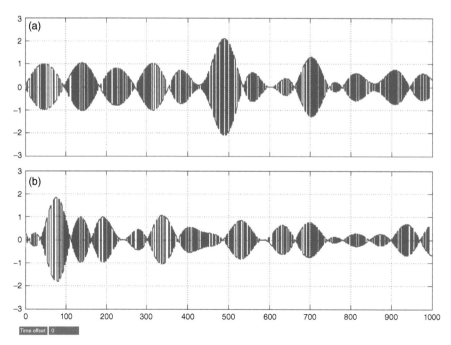

Figure 6.4 Real (a) and Imaginary (b) Parts of the Rayleigh Channel Output.

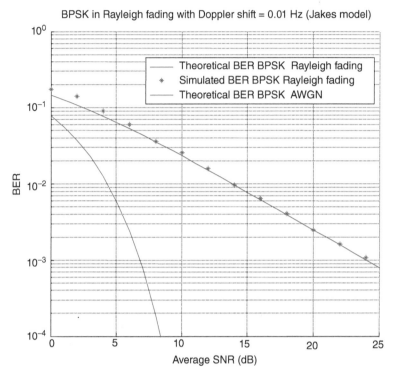

Figure 6.5 Theoretical and Simulated BPSK BER in Rayleigh Fading.

selected to be Rayleigh. Good agreement is observed between the simulated and theoretical results for BPSK BER in Rayleigh fading. The BER penalty associated with the presence of a fading channel vis-à-vis AWGN is obvious.

6.3 BPSK BER PERFORMANCE IN RICIAN FADING

Prior to developing BER results for BPSK in Rician fading, it is worthwhile investigating the characteristics of the Rician channel using the simple model in Figure 6.6 where the BPSK modulator is fed with random integers. In this model, the box for selection of Rician fading channel parameters is checked to open the channel visualization at the start of the simulation. In this simulation, specific parameters include the Rician K factor that specifies the ratio of the specular component to the diffuse multipath components, the Doppler shift associated with the specular (line-of-sight) component, and the maximum Doppler shift associated with the diffuse multipath components. This

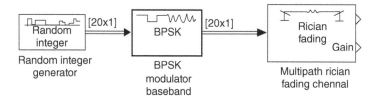

Figure 6.6 Simulink Model for Rician Channel.

simulation is a degenerate case where no multipath exists resulting in spectral response that exhibits flat fading.

A summary of the model parameters is specified as follows:

Model Parameters for BPSK in Rician Fading

- Antipodal signals = +1 and −1, 1 bit/symbol
- Sample time = symbol time = 1 s
- Frame based with 20 samples/frame
- Simulation time = 10,000 s
- Random integer seed = 22
- Jakes fading model with Doppler shift = 0.01 Hz
- Maximum diffuse Doppler shift = 0.1 Hz
- K factor = 3

When this simulation is executed, the channel visualization capability enables the selection of various results. Figures 6.7 and 6.8 display the impulse response and Doppler spectrum, respectively, after 10,000 s for the conditions specified earlier. Figure 6.8 shows the theoretical U-shaped, Jakes spectrum along with data samples indicating that the simulated spectrum is a good estimate.

Appendix 6.B presents a brief summary of BPSK BER performance in Rician fading. The Simulink model for this case is provided in Figure 6.9.

A summary of the model parameters is specified as follows:

Model Parameters for BPSK in Rician Fading

- Antipodal signals = +1 and −1, 1 bit/symbol
- Sample time = symbol time = 1 s

- Frame based with 20 samples/frame
- Simulation time = 1,000,000 s
- Random integer seed = 22
- Jakes fading model with Doppler shift = 0.01 Hz
- Maximum diffuse Doppler shift = 0.1 Hz
- K factor = 3
- Input signal power = 1 W
- Average SNR $\overline{\gamma_b}$ = 10 dB => Simulated BER = 0.008

Figure 6.10 displays the Rician channel parameter selection. Multipath parameters are not specified in this simulation since flat fading is assumed.

Figure 6.11 depicts BER results for BPSK in Rician fading for selected Rice factor values, $K = 1, 2, 3, 5, 10$.

For comparison, the theoretical BER for Rayleigh fading is included in Figure 6.11. Note that increasing the value of the Rice factor suggests a stronger dominant received signal component, thus resulting in an improved

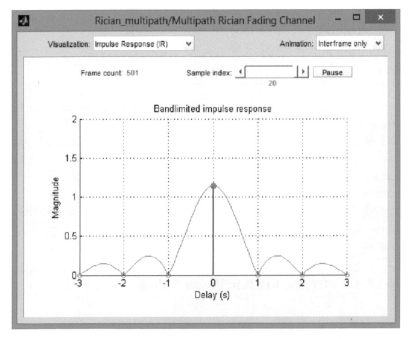

Figure 6.7 Rician Channel Impulse Response ($K = 3$).

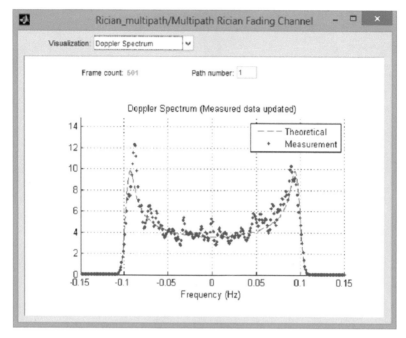

Figure 6.8 Rician Channel Doppler Spectrum ($K = 3$).

BER. Good agreement is observed between theoretical and simulated BER for all cases.

6.4 BPSK BER PERFORMANCE IN RICIAN FADING WITH MULTIPATH

The power of Simulink is realized here when multipath occurs and theoretical results are unavailable. Before computing BPSK BER results for Rician fading, the characteristics of the Rician channel will be reviewed.

Figure 6.12 shows the selection of the Rician channel parameters together with multipath. Introducing added parameters allows the insertion of multipath as a vector of path delays and a vector of average path gains. Here two paths are introduced with a main path and a second delayed path that has an average path gain defined as X. In this example, the main path has zero delay with an average path gain of 0 dB and the multipath component has a delay of 2 s with an average path gain, $X = -3$ dB. The average overall path gain is normalized to 0 dB.

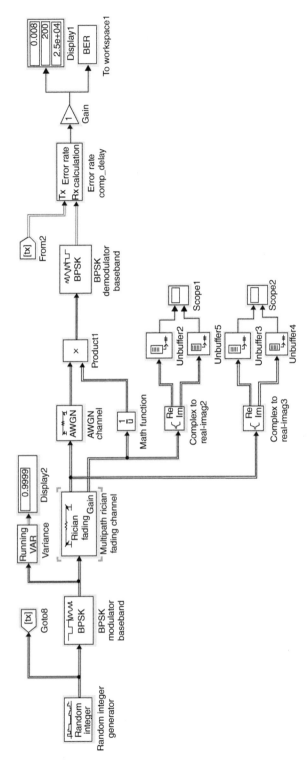

Figure 6.9 Simulation Model for Estimation of BPSK BER in Rician Fading.

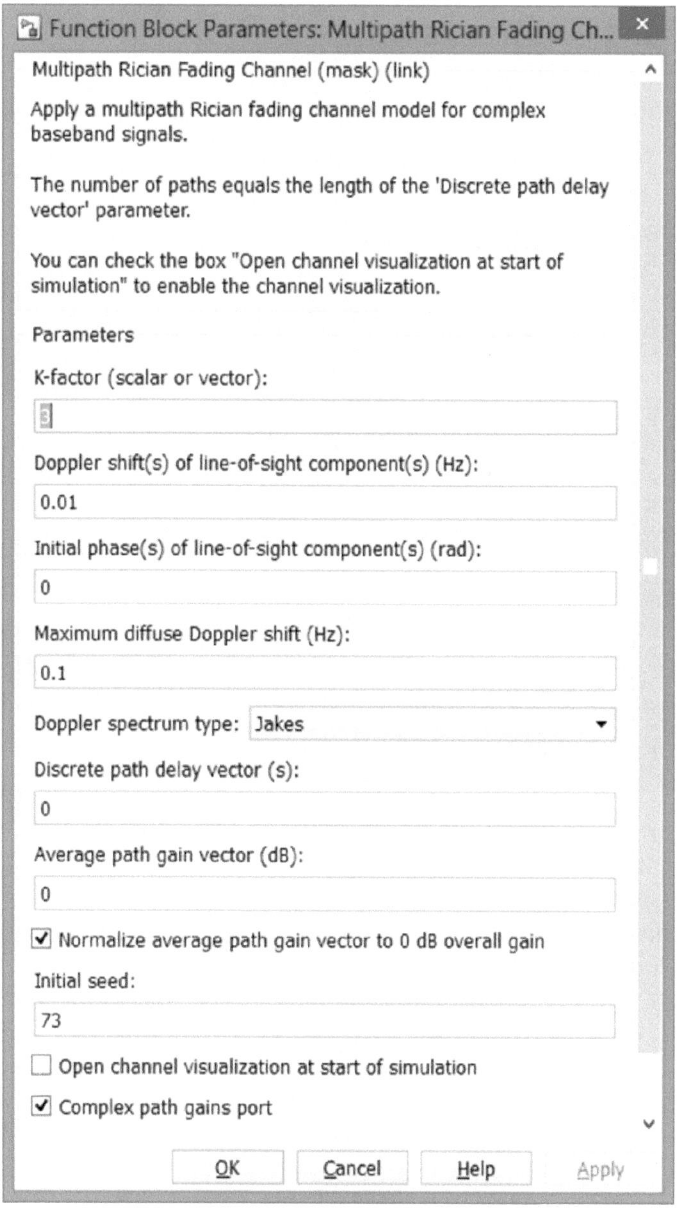

Figure 6.10 Rician Channel Parameters.

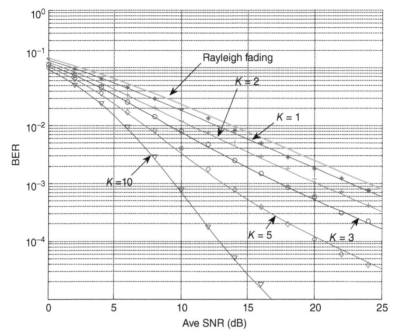

Figure 6.11 Simulated and Theoretical BER for BPSK in Rician Fading (solid lines are theoretical; simulated BER indicated with markers).

Using the Simulink model in Figure 6.6, the Rician fading channel characteristics with multipath can be investigated. Figures 6.13, 6.14, and 6.15 depict, respectively, the channel frequency response, the channel impulse response, and the Doppler spectrum. Figure 6.13a–c shows the variation in the frequency response due to the time variability of the channel with snapshots taken at 1000, 3140, and 10,000 s. It is observed that a null in the frequency response, resulting from the presence of multipath, occurs at 0.25 Hz from the peak. Figure 6.14a, b, and c show the time varying behavior of the impulse response as the simulation is stopped after 1000, 3140, and 10,000 s, respectively.

Comparing Figure 6.15 and Figure 6.8, it can be observed that the time-varying nature of the channel results in a poorer estimate of the Doppler spectrum, as shown in Figure 6.15.

To continue with BER estimation, the Simulink model for BPSK BER performance with multipath interference in a Rician fading channel is depicted in Figure 6.16. In the simulation, the math function block $1/u$ serves as an AGC by selecting the first arriving component to track the signal variation.

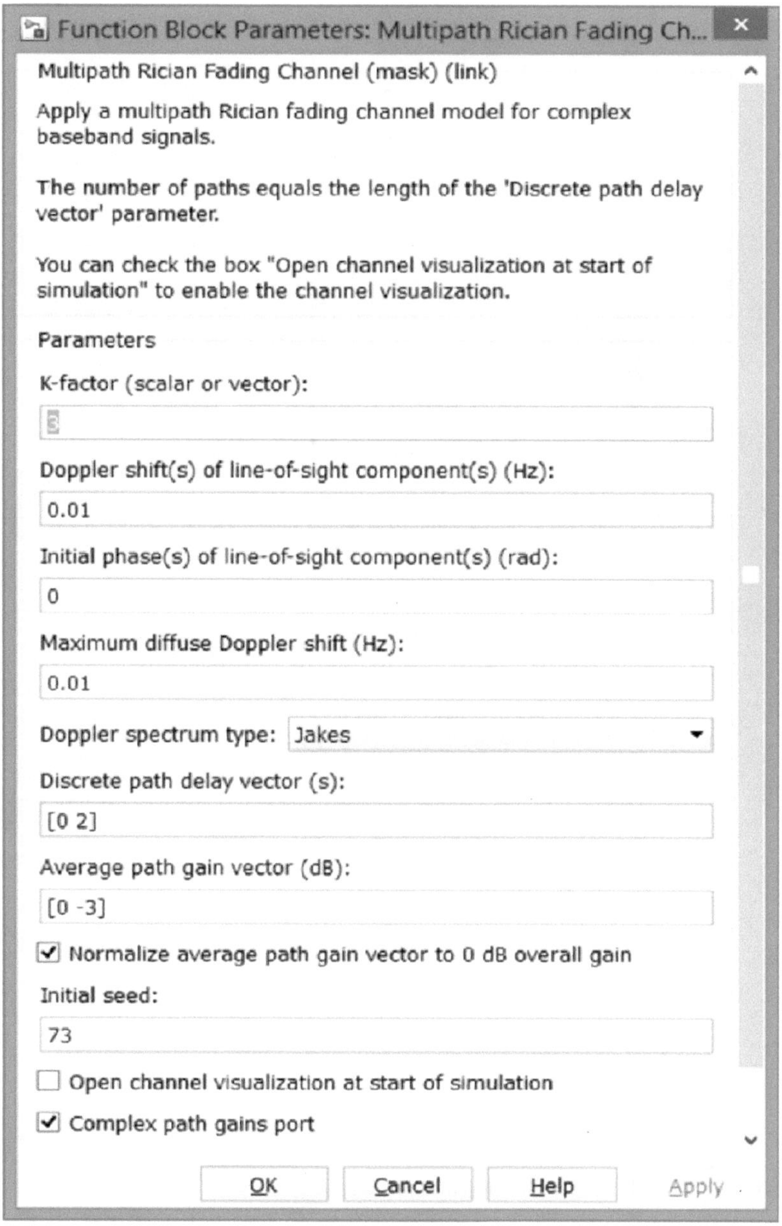

Figure 6.12 Rician Channel Parameters with Multipath.

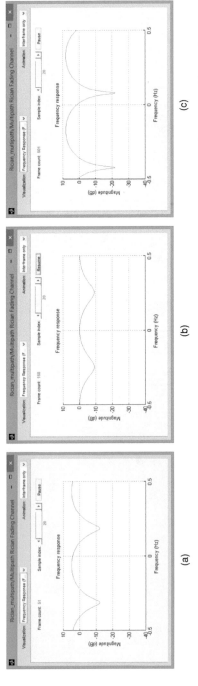

Figure 6.13 (a) Rician Channel Frequency Response: Multipath ($K=3$, $X=-3$ dB) after 1000 s; (b) Rician Channel Frequency Response: Multipath ($K=3$, $X=-3$ dB) after 3140 s; (c) Rician Channel Frequency Response:Multipath ($K=3$, $X=-3$ dB) after 10,000 s.

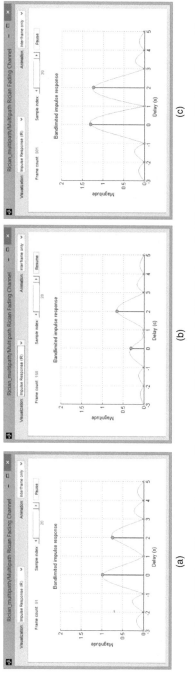

Figure 6.14 (a) Rician Channel Impulse Response with Multipath ($K = 3$, $X = -3$ dB) after 1000 s; (b) Rician Channel Impulse Response with Multipath ($K = 3$, $X = -3$ dB) after 3140 s; (c) Rician Channel Impulse Response with Multipath ($K = 3$, $X = -3$ dB) after 10,000 s.

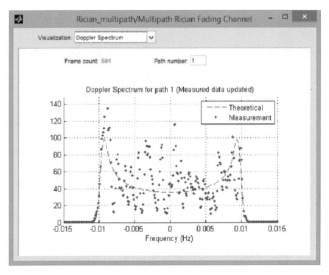

Figure 6.15 Rician Channel Doppler Spectrum ($K = 3$, $X = -3$ dB).

A summary of the model parameters is specified as follows:

Model Parameters for BPSK in Rician Fading with Multipath

- Antipodal signals = +1 and −1, 1 bit/symbol
- Sample time = symbol time = 1 s
- Sample based
- Simulation time = 10,000 s
- Random integer seed = 22
- Jakes fading model with Doppler shift = 0.01 Hz
- Maximum diffuse Doppler shift = 0.01 Hz
- K factor = 3
- Path delay vector = [0 2] s
- Average path gain vector = [0 −3] dB, 0 dB overall gain
- Input signal power = 1 W
- Average SNR $\overline{\gamma_b}$ = 10 dB => Simulated BER = 0.056

The BER = 0.056 shown in Figure 6.16 is obtained using the parameters identified in the box shown above and causes a reduction in BER once

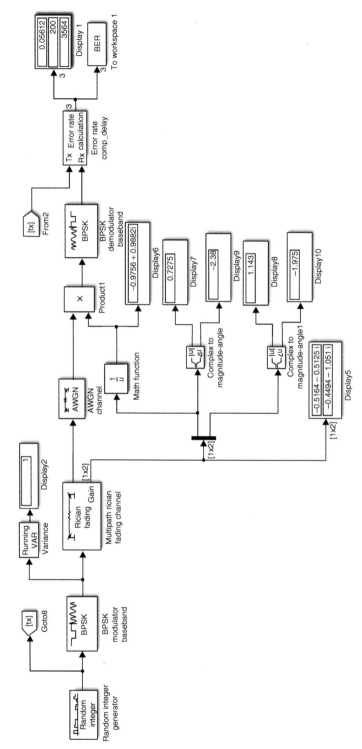

Figure 6.16 Model for Estimation of BPSK BER in a Rician Fading Channel with Multipath.

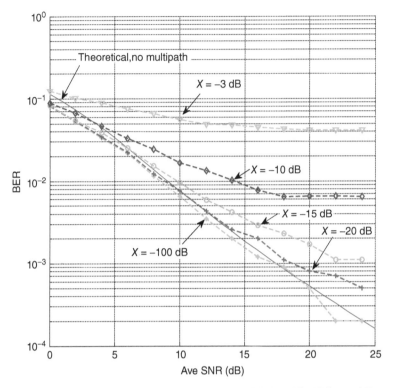

Figure 6.17 BPSK BER in Rician Fading with Specific Values of X.

multipath is included. To confirm this result, Figure 6.16 is modified to set the maximum diffuse Doppler shift to 0.1 Hz and the average path gain of the delayed component to $X = -100$ dB. The error probability is then seen to be 0.008 with an average SNR $\overline{\gamma_b} = 10$ dB; this BER is in agreement with the BER obtained in the Figure 6.6 Simulink model where no multipath is specified.

The model shown in Figure 6.16 is now re-executed with the maximum diffuse Doppler shift retained at 0.1 Hz. Changing the second multipath component gain to $X = -3$ dB results in BER = 0.08; this BER can be compared to the case cited earlier with no multipath where BER = 0.008. The interference introduced by the multipath is seen to cause a severe increase in BER even when the second multipath component is 3 dB lower than the main path.

The model shown in Figure 6.16 is now modified to investigate the effect associated with variations in the second average path gain, X. Using the bertool, Figure 6.17 provides BPSK BER results for Rician fading channels versus X in dB. The model parameters are specified as follows:

> **Model Parameters for BPSK in Rician Fading with Multipath**
>
> - Antipodal signals = +1 and −1, 1 bit/symbol
> - Sample time = symbol time = 1 s
> - Sample based
> - Simulation time = 10,000 s
> - Random integer seed = 22
> - Jakes fading model with Doppler shift = 0.01 Hz
> - Maximum diffuse Doppler shift = 0.01 Hz
> - K factor = 3
> - Path delay vector = [0 2] s
> - Average path gain vector = [0 X] dB, 0 dB overall gain
> - Input signal power = 1 W

Figure 6.17 shows that with a low-value of X, that is, $X = -100$ dB, the BER performance agrees with the theoretical BER for the no-multipath case. As the multipath gain X gets larger, significant loss in BER occurs.

6.5 SUMMARY DISCUSSION

This chapter has demonstrated the application of commonly used channel fading models available in Simulink. When the channel exhibits fading, BER performance for specific modulations depends on the selected fading model and is severely degraded from performance in AWGN. Examples provided here included Rayleigh and Rican fading models. Simulink models allow the BER performance in the presence of multipath to be readily estimated. This capability provided by Simulink is important since estimation of BER in the presence of multipath is not easily obtained analytically.

APPENDIX 6.A THEORETICAL BER PERFORMANCE OF BPSK IN RAYLEIGH FADING

In Appendix 3.A, the received BPSK signal is represented by

$$r(t) = \alpha e^{-j\emptyset} u_i(t) + z(t), \ 0 \le t \le T, \ i = 1, 2$$

In fading the parameter α is no longer a constant and in the case of frequency nonselective slow fading is a Rayleigh-distributed, random variable.

The Rayleigh fading behavior exists when a signal that is received over multiple reflective paths has no dominant component. The assumption of slow fading implies that the channel impulse response exhibits a slower variation than the rate of change of the transmitted signal thereby allowing the phase shift $\emptyset$ to be precisely estimated from the received signal without error.

In Appendix 3.A, the energy contrast ratio is expressed as $\gamma_b = \alpha^2 \frac{E_b}{N_o}$. Since α is random, the energy contrast ratio is also random with an average value $\overline{\gamma_b} = \frac{E_b}{N_o} E(\alpha^2)$. Assuming X_1 and X_2 are independent, zero mean Gaussian random variables each with variance β^2, α^2 can be expressed as $\alpha^2 = X_1^2 + X_2^2$ and is observed to be chi-squared with 2 degrees of freedom. Moreover, the probability density function (pdf) of γ_b is then found to be

$$p(\gamma_b) = \frac{1}{\overline{\gamma_b}} e^{-\frac{\gamma_b}{\overline{\gamma_b}}}, \quad \gamma_b \geq 0$$

with $\overline{\gamma_b} = \frac{E_b}{N_o} E(\alpha^2) = \frac{E_b}{N_o}(2\beta^2)$. The error probability, P_b, for BPSK in Rayleigh fading is found by averaging $P_o = \frac{1}{2}\text{erfc}(\sqrt{\gamma_b})$ over the pdf of γ_b, that is,

$$P_b = \int_0^\infty \frac{1}{2}\text{erfc}(\sqrt{\gamma_b}) \frac{1}{\overline{\gamma_b}} e^{-\frac{\gamma_b}{\overline{\gamma_b}}} d\gamma_b$$

Performing the integration results in

$$P_b = \frac{1}{2}\left[1 - \sqrt{\frac{\overline{\gamma_b}}{1+\overline{\gamma_b}}}\right]$$

where $\overline{\gamma_b}$ is the average SNR/bit.[2]

APPENDIX 6.B THEORETICAL BER PERFORMANCE OF BPSK IN RICIAN FADING

A Rician channel can be viewed as a generalization of Rayleigh fading where a strong dominant component occurs along with multiple reflective paths. It is again assumed that the channel exhibits slow fading allowing the phase shift $\emptyset$ to be estimated without error. The parameter α is now characterized as a Rician-distributed random variable. Assuming X_1 and X_2 are independent, Gaussian random variables each with mean μ and variance β^2, α^2 expressed as

---

[2] Proakis J.G., and M. Salehi, Digital Communications, 5th ed pp. 846–849.

$\alpha^2 = X_1^2 + X_2^2$ is seen to be noncentral chi-squared with 2 degrees of freedom. The energy contrast ratio $\gamma_b = \alpha^2 \frac{E_b}{N_o}$ now has an average value given by

$$\overline{\gamma_b} = \frac{E_b}{N_o} E(\alpha^2) = \frac{E_b}{N_o}(2\mu^2 + 2\beta^2)$$

It is convenient to define a Rice factor, $K = \frac{\text{specular component}}{\text{random component}}$ or $K = \frac{\mu^2}{\beta^2}$. The pdf of γ_b is then found to be[3]

$$p(\gamma_b) = \frac{1+K}{\overline{\gamma_b}} e^{-\frac{\gamma_b(1+K) + K\overline{\gamma_b}}{\overline{\gamma_b}}} I_0\left(\sqrt{\frac{4(1+K)K\gamma_b}{\overline{\gamma_b}}}\right), \quad \gamma_b \geq 0$$

The probability of error, obtained by averaging P_o over the pdf of γ_b, is then given by[4]

$$P_b = Q(a,b) - \frac{1}{2}\left[1 + \sqrt{\frac{\overline{\gamma_b}}{1+\overline{\gamma_b}}}\right] e^{-(a^2+b^2)/2} I_0(ab)$$

where $Q(a,b) = \int_b^\infty x e^{-(x^2+a^2)/2} I_0(ax) dx$ and

$$a = \sqrt{\frac{K[1 + 2\overline{\gamma_b} - 2\sqrt{\overline{\gamma_b}(1+\overline{\gamma_b})}]}{2[1+\overline{\gamma_b}]}}$$

$$b = \sqrt{\frac{K[1 + 2\overline{\gamma_b} + 2\sqrt{\overline{\gamma_b}(1+\overline{\gamma_b})}]}{2[1+\overline{\gamma_b}]}}$$

PROBLEMS

6.1 In the Figure 6.16 Simulink model, verify that the output of the math function block 1/u is correct.

6.2 Retaining $\overline{\gamma_b} = 10$ dB, and the Jakes model with Doppler shift = 0.01 Hz, modify the Figure 6.16 Simulink model to use 0.1 Hz as the

---

[3] Rappaport, T., Wireless Communications, Principles and Practice, Prentice Hall, 1966, p. 288.
[4] Lindsey, W., Error probabilities for Rician Fading Multichannel Reception of Binary and N-ary Signals, IEEE Trans on Information Theory, Oct 1964, pp. 339–350.

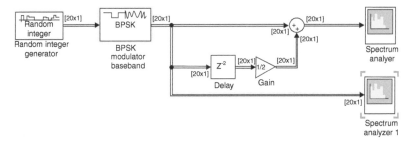

Figure P.6.1   Multipath Spectrum.

maximum diffuse Doppler shift with the average second path gain $X = -100$ dB.

a. What is the resulting BER?

b. Repeat this computation for $X = -3$ dB and determine the BER.

6.3 Figure 6.16 shows that at the end of the 10,000 s simulation, the magnitude of the main multipath component is 0.7275 and the second component is 1.143. Using a modified simulation based on the same Rician and AWGN parameter selections, show that over the 10,000 s simulation time, the main component magnitude is on average greater than the second component magnitude. What is the magnitude of these components after 10,000 s in your simulation?

6.4 Continuing with the simulation developed in Prob 6.3 change the model to use the second path for the AGC. What is the resulting BER and how do you explain the change?

6.5 Prove that the null exhibited in Figure 6.13 occurs at 0.25 Hz from the peak.

6.6 Using the Simulink model provided in Figure P.6.1, compute the spectrum with the parameters RBW = 0.05 Hz, trace options = 100 average and frequency options = linear, 50% overlap and rectangular window

a. Find the frequency null and explain its location.

b. Determine and explain the magnitude of the normalized spectrum from Prob 6.5 at zero frequency.

c. Increase the delay to 4 s and determine the location of the null.

7

DIGITAL COMMUNICATIONS BER PERFORMANCE IN AWGN (FSK IN FADING)

7.1 FSK IN RAYLEIGH AND RICIAN FADING

This chapter discusses several topics in Simulink based on FSK modulation in fading channels. Specifically these topics include the following:

- BFSK BER performance in Rayleigh fading
- MFSK BER performance in Rayleigh fading
- BFSK BER performance in Rician fading
- CPFSK BER performance in Rician fading with Multipath

7.2 BFSK BER PERFORMANCE IN RAYLEIGH FADING

This section discusses BFSK BER performance in Rayleigh fading. A brief review of the theoretical BFSK BER performance for Rayleigh fading channels is provided in Appendix 7.A. The error probability for BFSK in Rayleigh fading assuming noncoherent detection is given by

$$P_b = \frac{1}{2 + \overline{\gamma_b}}$$

Modeling of Digital Communication Systems Using SIMULINK®, First Edition.
Arthur A. Giordano and Allen H. Levesque.
© 2015 John Wiley & Sons, Inc. Published 2015 by John Wiley & Sons, Inc.
Companion Website: www.wiley.com/go/simulink

where $\overline{\gamma_b}$ is the average SNR/bit. The time-varying nature of the channel is again characterized by its power spectral density $S(f)$ according to the Jakes fading model presented in Chapter 6.

A simulation to estimate the BER for BFSK with noncoherent detection in Rayleigh fading is shown in Figure 7.1.

A summary of the model parameters is specified as follows:

Model Parameters for BFSK in Rayleigh Fading

- Binary orthogonal signals, 1 bit/symbol
- Frame based with 1 sample/frame
- Symbol period = 0.2 s
- 1000 samples/symbol
- FSK tones = ±50 Hz with 100 Hz separation
- Simulation time = 1500 s
- Bernoulli binary seed = 61
- Jakes fading model with Doppler shift = 0.5 Hz
- Input signal power = 1 W
- Average SNR = 10 dB =>
  - Simulated BER = 0.085
  - Theoretical BER = 0.083

An illustration of the BFSK power spectral density at the Rayleigh channel output is shown in Figure 7.2 where the tones appear at ±50 Hz. This plot was obtained from the spectrum analyzer with scope settings: 7.5 Hz RBW, 100 averages with 1000 point FFT and a Hann window with no overlap. Figure 7.3 displays a 10 s snapshot of the Rayleigh channel output where signal fading is clearly visible.

Figure 7.4 shows the BER performance of BFSK with noncoherent detection in Rayleigh fading obtained by executing the simulation as shown in Figure 7.1 with the bertool. Good agreement is observed between the theoretical and simulated BER performance. For reference, the theoretical BFSK BER in AWGN is shown where the penalty due to fading is found to be significant.

7.3 MFSK BER PERFORMANCE IN RAYLEIGH FADING

Next, M-ary FSK BER performance in Rayleigh fading is presented. The theoretical result for the probability of error P_b is summarized in Appendix 7.A.

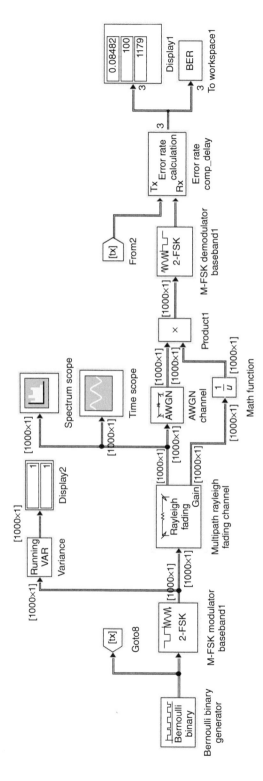

Figure 7.1 Simulink Model for Estimation of BFSK BER in Rayleigh Fading.

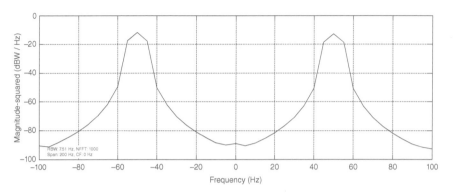

Figure 7.2 BFSK Power Spectral Density at Rayleigh Channel Output.

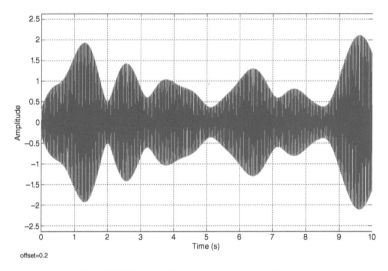

Figure 7.3 BFSK Time Display at Rayleigh Channel Output.

for $M = 2^k$ and is given here as

$$P_b = \frac{2^{k-1}}{2^k - 1} \sum_{i=1}^{M-1} \frac{-1^{i-1} \binom{M-1}{i}}{1 + i + i\overline{\gamma}_c}$$

where $\overline{\gamma}_b = \frac{\overline{\gamma}_c}{k}$. Figure 7.5 shows the Simulink model used to compute the BER performance of 32 FSK BER in Rayleigh fading with noncoherent detection.

MFSK BER PERFORMANCE IN RAYLEIGH FADING

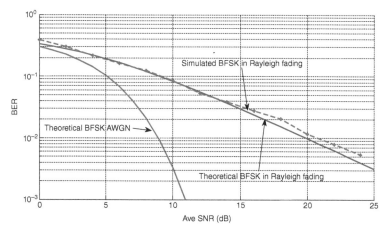

Figure 7.4 BFSK BER in Rayleigh Fading.

In the Simulink model $k = \log_2 32 = 5$, the gain $K = 16/31$, and $\overline{\gamma_b} = 10$ dB. A summary of the model parameters is provided as follows:

Model Parameters for MFSK in Rayleigh Fading

- $M = 32$-ary orthogonal signals, 1 bit/symbol, $k = 5$
- Frame based with 1 sample/frame
- Symbol period = 0.2 s
- 1000 samples/symbol
- 100 Hz frequency separation
- Simulation time = 1500 s
- Random integer seed = 32
- Jakes fading model with Doppler shift = 0.5 Hz
- Input signal power = 1 W
- $E_s/N_o = 10 + 10*\log(5)$
- Average SNR = 10 dB =>
  - Simulated BER = 0.037 (stop after 100 errors)

Executing the Simulink model shown in Figure 7.5 results in the BER performance for MFSK in Rayleigh fading with noncoherent detection. Figure 7.6 plots the BER performance for $M = 2$, 4, and 32. In Figure 7.6 the solid lines represent theoretical performance and the markers indicate simulated performance.

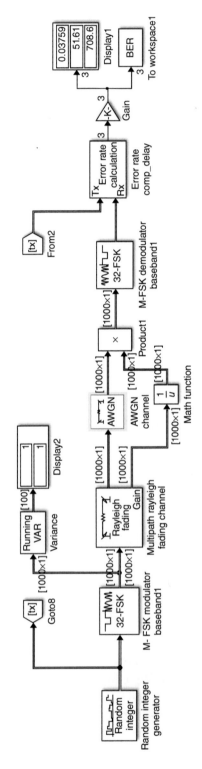

Figure 7.5 Simulink Model for Estimation of MFSK BER in Rayleigh Fading.

146

BFSK BER PERFORMANCE IN RICIAN FADING

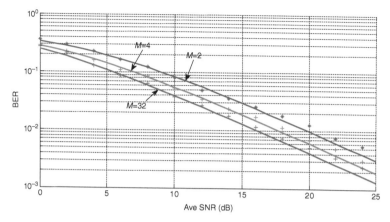

Figure 7.6 BER performance for MFSK in Rayleigh fading.

7.4 BFSK BER PERFORMANCE IN RICIAN FADING

The theoretical BER performance of BFSK in Rician fading with noncoherent detection is discussed in Appendix 7.A; the probability of error, P_b, is given by

$$P_b = \frac{1+K}{2+2K+\overline{\gamma_b}} e^{-\left(\frac{K\overline{\gamma_b}}{2+2K+\overline{\gamma_b}}\right)}$$

A Simulink model for computing this BER is shown in Figure 7.7. The model parameters are specified as follows:

Model Parameters for BFSK in Rician Fading

- Binary orthogonal signals, 1 bit/symbol
- Frame based with 1 sample/frame
- Symbol period = 0.2 s
- 1000 samples/symbol
- FSK tones = ±50 Hz with 100 Hz separation
- Simulation time = 1500 s
- Bernoulli binary seed = 61
- Jakes fading model with $K = 3$
- Maximum diffuse Doppler shift = 0.5 Hz
- No multipath

- Input signal power = 1 W
- Average SNR = 10 dB =>
  - Simulated BER = 0.04
  - Theoretical BER = 0.042

A comparison of the simulated and theoretical BER performance for BFSK in Rician fading with noncoherent detection is provided in Figure 7.8.

7.5 BFSK BER PERFORMANCE IN RICIAN FADING WITH MULTIPATH

Figure 7.9 shows the Simulink model for computing BER performance in Rician fading where multipath parameters are included. The modulation is CPFSK with a modulation index, $h = 0.75$, and is coherently detected. The model shown in Figure 7.9 follows the implementation from Chapter 6 where the math function block $1/u$ serves as an AGC by selecting the first arriving component to track the signal variation. This implementation allows the effects associated with variations in the second average path gain, X to be investigated. The model parameters are as follows:

Model Parameters for CPFSK in Rician Fading with multipath

- Binary CPFSK coherent detection, $h = 0.75$
- Sample based with sample time = 1 s
- Symbol period = 1 s
- 1 sample/symbol
- 1 bit/symbol
- Simulation time = 15,000 s
- Random integer seed = 37
- Jakes fading model with $K = 3$
- Maximum diffuse Doppler shift = 0.1 Hz
- Path delay vector = [0 2] s
- Traceback depth = 16
- Average path gain vector = [0 −100] dB
- Input signal power = 1W
- Average SNR = 20 dB =>
  - Simulated BER = 0.0017

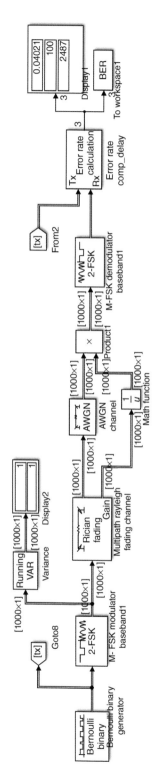

Figure 7.7 Simulink Model for Estimation of BFSK BER in Ricean Fading.

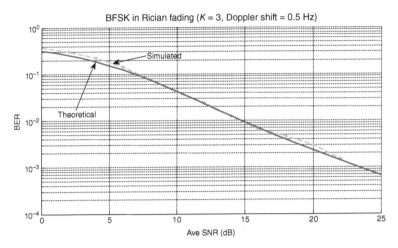

Figure 7.8 BER for BFSK in Rician Fading.

Using the bertool, Figure 7.10 shows CPFSK BER results for Rician fading channels with multipath versus X in dB. For comparison with the simulated results, theoretical BERs for coherently detected FSK in Rayleigh fading and Rician fading with $K = 1$ are depicted. All of the simulations correspond to coherently detected CPFSK. The simulations with $X = -100$ dB represent cases with no multipath. It is clear that even with relatively weak multipath, BER performance is severely degraded.

7.6 SUMMARY DISCUSSION

Theoretical FSK BER performance in AWGN is well known. However, when the channel exhibits fading, BER performance for specific modulations depends on the selected fading model and is severely degraded from performance in AWGN. Examples provided here included Rayleigh and Rican fading models. Simulink models allow the BER performance in the presence of multipath to be readily estimated. This capability, provided by Simulink, is important since estimation of BER in the presence of multipath is not easily obtained analytically.

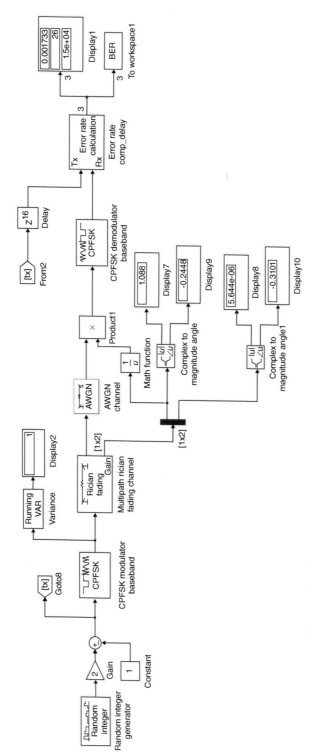

Figure 7.9 Simulink Model for Estimation of CPFSK BER in Rician Fading with Multipath.

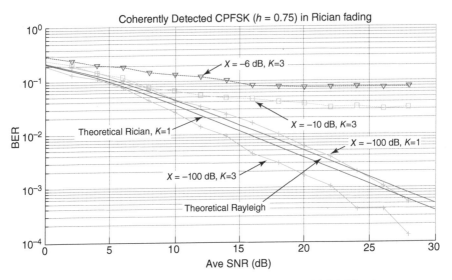

Figure 7.10 CPFSK BER in Rician Fading with Multipath.

APPENDIX 7.A THEORETICAL BER PERFORMANCE OF FSK IN RAYLEIGH AND RICIAN FADING

Rayleigh Fading

The error probability, P_2, of binary FSK with orthogonal signals in AWGN is obtained assuming noncoherent detection; P_2 is then given by

$$P_2 = \frac{1}{2} e^{-\frac{\gamma_b}{2}}$$

where $\gamma_b = \alpha^2 \frac{E_b}{N_o}$. In the presence of frequency nonselective slow fading, the parameter α is a Rayleigh-distributed, random variable due to the reception of multiple reflective paths with no dominant component. The parameter α^2 is then chi-squared with 2 degrees of freedom. From Appendix 6.A, the pdf of γ_b is

$$p(\gamma_b) = \frac{1}{\overline{\gamma_b}} e^{-\frac{\gamma_b}{\overline{\gamma_b}}}, \quad \gamma_b \geq 0$$

where $\overline{\gamma_b} = \frac{E_b}{N_o} E(\alpha^2)$.

SUMMARY DISCUSSION

The error probability for BFSK in Rayleigh fading with noncoherent detection is then found by averaging $P_2 = \frac{1}{2}e^{-\frac{\gamma_b}{2}}$ over the pdf of γ_b, resulting in[1]

$$P_b = \frac{1}{2 + \overline{\gamma_b}}$$

The error probability for MFSK in Rayleigh fading with noncoherent detection is known and is given by[1]

$$P_b = \frac{2^{k-1}}{2^k - 1} \sum_{i=1}^{M-1} \frac{-1^{i-1} \binom{M-1}{i}}{1 + i + i\overline{\gamma_c}}$$

where

$$\overline{\gamma_b} = \frac{\overline{\gamma_c}}{k}, k = \log_2 M$$

Rician Fading

From Appendix 6.B, α^2 is known to be non-central chi-squared with 2 degrees of freedom. In Rician fading, the pdf of γ_b is then found to be

$$p(\gamma_b) = \frac{1+K}{\overline{\gamma_b}} e^{-\frac{\gamma_b(1+K) + K\overline{\gamma_b}}{\overline{\gamma_b}}} I_0\left(\sqrt{\frac{4(1+K)K\gamma_b}{\overline{\gamma_b}}}\right), \gamma_b \geq 0$$

where K is the Rice factor.

The probability of error for BFSK in Rician fading with noncoherent detection, obtained by averaging P_2 over the pdf of γ_b, is then given by[2]

$$P_b = \frac{1+K}{2 + 2K + \overline{\gamma_b}} e^{-\left(\frac{K\overline{\gamma_b}}{2+2K+\overline{\gamma_b}}\right)}$$

---

[1] Proakis, J.G., and M. Salehi, Digital Communications, 5th ed, pp. 847–849.
[2] Rappaport, T., Wireless Communications, Principles and Practice, Prentice Hall, 1966, p. 288.

PROBLEMS

7.1 Change the BFSK tone spacing to 500 Hz and obtain the simulated BER performance for BFSK in Rayleigh fading assuming noncoherent detection over the range 0 to 25 dB in 2 dB steps.

  a. List the model parameters.

  b. How does this result compare with theoretical performance?

7.2 Using the model in Figure P.7.1 with the model parameters used for Figure 7.4, obtain the simulated and theoretical BER performance for BFSK in Rayleigh fading assuming noncoherent detection with a symbol time of 2 s. Explain the result.

7.3 Execute the model in Figure P.7.1 for binary CPFSK in AWGN using the following model parameters:

Model Parameters for CPFSK in AWGN (See Figure P.7.1)

- 1 bit/symbol, 1 sample/symbol
- Sample based
- Symbol period = 1 s
- Modulation index = 0.75
- Simulation time = 1500 s
- Random integer seed = 37
- Input signal power = 1 W
- Traceback depth = 16

What is the BER for $\gamma_b = 6$ dB?

For a range of values $\gamma_b = 0, 2, 4, 6, 8, 10$ dB obtain the simulated BER and compare it with the theoretical BER.

Replace the AWGN channel with a Rayleigh channel having no multipath and a maximum Doppler shift = 0.1 Hz. For $\overline{\gamma}_b$ in 2 dB steps between 0 and 20 and $h = 0.75$, and compare the theoretical and simulated BER. Repeat the simulated result for $h = 0.7$.

7.4 For $K = 3$ and $\overline{\gamma}_b = 10$ dB, compute the theoretical BER for BFSK in Rician fading with noncoherent detection. And discuss how it compares with the simulated BER.

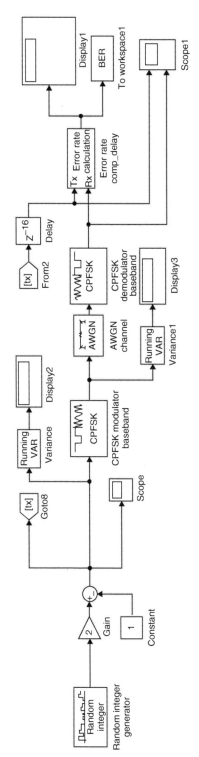

Figure P.7.1 Model for Estimation of CPFSK BER in AWGN.

7.5 Compute the simulated and theoretical BER performance for BFSK in Rician fading for $K = 1, 3,$ and 10 with no multipath and a maximum diffuse Doppler shift $= 0.5$ Hz. Assume $\overline{\gamma_b}$ has 2 dB steps; for $K = 1$ let the maximum value of $\overline{\gamma_b} = 30$, for $K = 3$ the maximum value of $\overline{\gamma_b} = 26$ and for $K = 10$ the maximum value of $\overline{\gamma_b} = 16$.

8

DIGITAL COMMUNICATIONS BER PERFORMANCE (STBC)

8.1 DIGITAL MODULATIONS IN RAYLEIGH FADING WITH STBC

This chapter introduces several topics in Simulink using space–time block coding (STBC) for compensating for BER degradation due to channel fading. Specifically these topics include the following:

- BPSK in Rayleigh fading with STBC
- QAM in Rayleigh fading with STBC

8.2 BPSK BER IN RAYLEIGH FADING WITH STBC

Appendix 8.A presents an overview of Alamouti STBC for two transmit antennas and one receive antenna. The formulation can be generalized to N transmit antennas and M receive antennas as implemented in Matlabs's Communication System Toolbox. Figures 8.1 and 8.2 show a Simulink model for estimating the BPSK BER performance in Rayleigh fading with two transmit antennas and one receive antenna using Alamouti STBC.

Modeling of Digital Communication Systems Using SIMULINK®, First Edition.
Arthur A. Giordano and Allen H. Levesque.
© 2015 John Wiley & Sons, Inc. Published 2015 by John Wiley & Sons, Inc.
Companion Website: www.wiley.com/go/simulink

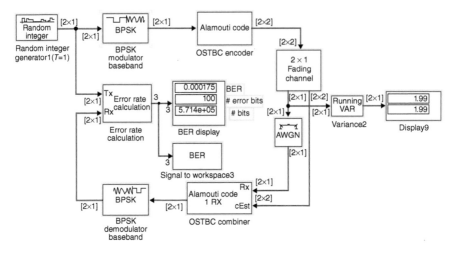

Figure 8.1 Simulink Model for BPSK BER in Rayleigh Fading with 2 × 1 STBC.

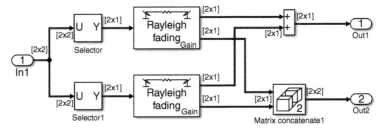

Figure 8.2 Rayleigh Fading Channel for Two Transmit Antennas and One Receive Antenna.

The parameters for this model are specified as follows:

Model Parameters for BPSK in Rayleigh Fading with STBC

- Antipodal signals = +1 and −1, 1 bit/symbol
- Sample time = symbol time = 1 s
- Frame based, 2 samples/frame
- Simulation time = stop with 100 errors
- Random integer seed = 22

- Jakes fading model with Doppler shift = 0.01 Hz for each channel
- Input signal power = 2 W
- Alamouti code: 2 trans,1 rcv antenna (diversity = 2)
- Average SNR = 18 dB =>
  - Simulated BER = 0.000175
  - Theoretical BER = 0.00019

Using the Simulink models shown in Figures 8.1 and 8.2 in conjunction with the bertool, the BPSK BER performance in Rayleigh fading with 2×1 STBC is obtained as illustrated in Figure 8.3. Using the BER formula in Appendix 8.A, theoretical results for diversity 1 and 2 are displayed for comparison with the simulated data.

Figures 8.4 and 8.5 show a Simulink model for estimating the BPSK BER performance in Rayleigh fading with two transmit antennas and two receive antennas using Alamouti STBC.

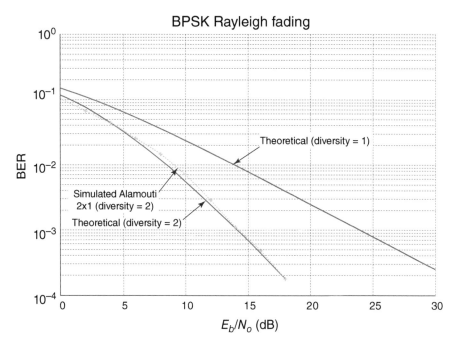

Figure 8.3 BPSK BER in Rayleigh Fading with Alamouti 2×1 STBC.

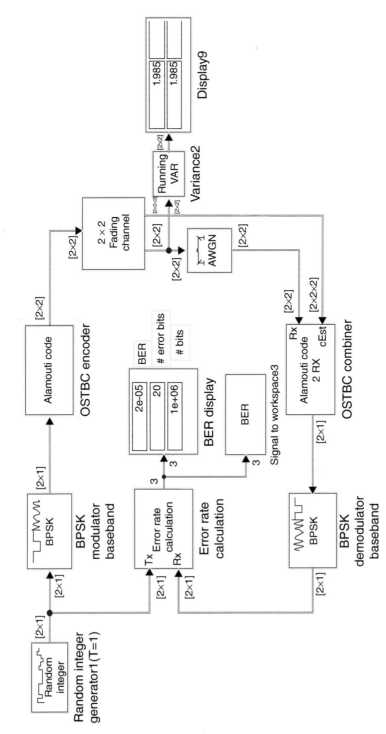

Figure 8.4 Simulink Model for BPSK BER in Rayleigh Fading with 2×2 STBC.

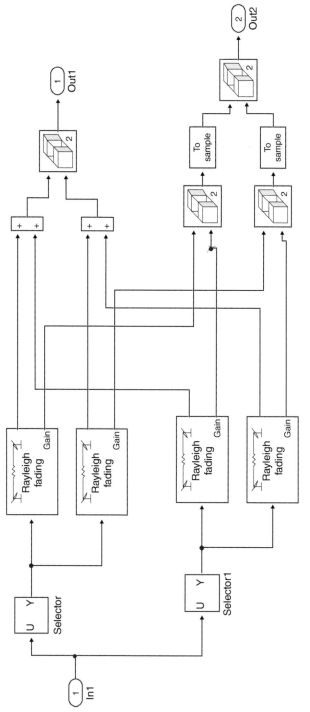

Figure 8.5 Rayleigh Fading Channel for 2 Transmit Antennas and 2 Receive Antennas.

The parameters for this model are specified as follows:

Model Parameters for BPSK in Rayleigh Fading with STBC

- Antipodal signals = +1 and −1, 1 bit/symbol
- Sample time = symbol time = 0.001 s
- Frame based, 2 samples/frame
- Simulation time = stop with 100 errors
- Random integer seed = 22
- Jakes fading model with Doppler shift = 3Hz for each channel
- Input signal power = 2 W
- Alamouti code: 2 trans, 2 rcv antennas (diversity = 4)
- Average SNR = 15 dB =>
  - Simulated BER = 2×10^{-5}

Figure 8.6 displays the BPSK BER in Rayleigh fading with 2×2 Alamouti STBC along with the theoretical results obtained with the formula in Appendix 8.A.

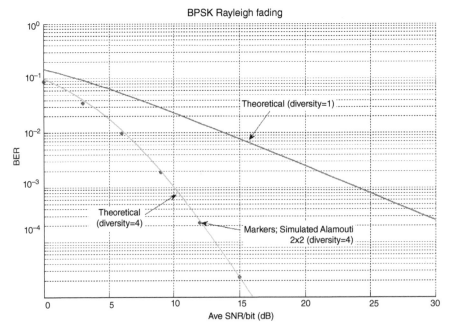

Figure 8.6 BPSK BER in Rayleigh Fading with 2×2 Alamouti STBC.

8.3 QAM BER IN RAYLEIGH FADING WITH STBC

BER performance for 16-QAM in Rayleigh fading and diversity is summarized in Appendix 8.B with two transmit antennas and L receive antennas. A Simulink model for two transmit antennas and two receive antennas is provided in Figure 8.7. The Rayleigh fading channel model is displayed in Figure 8.5.

The model parameters are specified as follows:

Model Parameters for 16-QAM in Rayleigh Fading with STBC

- Sample time = symbol time = 1 s
- Frame based, 2 samples/frame
- Simulation time = stop with 100 errors
- Random integer seed = 22
- Jakes fading model with Doppler shift = 0.01 Hz for each channel
- Input signal power = 2 W
- Alamouti code: 2 trans, 2 rcv antennas (diversity = 4)
- Average SNR = 10 dB =>
  - Simulated BER = 0.0081
  - Theoretical BER = 0.0083

The BER performance for 16-QAM in Rayleigh fading with 2 transmit antennas and 2 receive antennas is shown in Figure 8.8. It is observed that using 2×2 Alamouti STBC, the simulated BER performance agrees favorably with the theoretical performance obtained from Appendix 8.B.

8.4 SUMMARY DISCUSSION

Theoretical BPSK BER performance in AWGN is well known. However, when the channel exhibits fading, BER performance for specific modulations depends on the selected fading model and is severely degraded from performance in AWGN. Space–time block coding using multiple transmit and/or receive antennas can substantially improve BER performance in fading channels. Examples based on BPSK and 16-QAM modulations in Rayleigh fading with STBC have been presented that demonstrate diversity attained by use of multiple-antenna schemes.

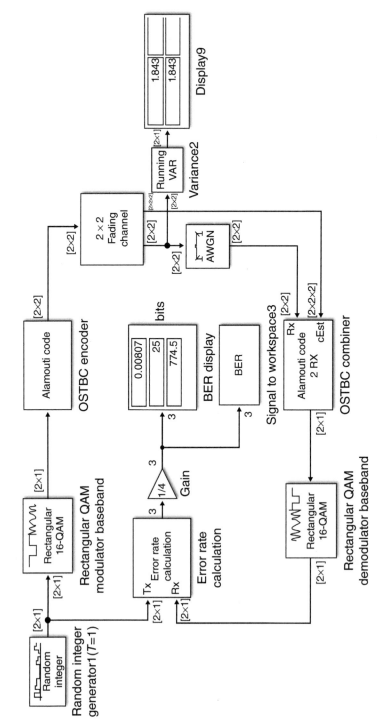

Figure 8.7 Simulink Model for 16-QAM BER in Rayleigh Fading with 2×2 Alamouti STBC.

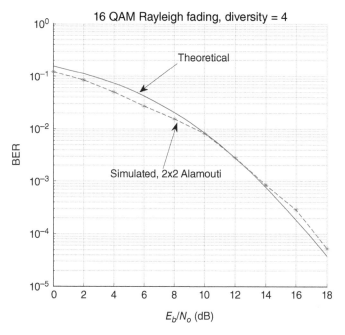

Figure 8.8 BER for 16 QAM in Rayleigh Fading with 2×2 Alamouti STBC.

APPENDIX 8.A SPACE–TIME BLOCK CODING FOR BPSK

Diversity techniques have been found to offer significant performance benefit for independent fading channels. Use of multiple transmit and/or multiple receive antennas (referred to as MIMO) for transmission over independent channels provides the diversity. Alamouti developed a scheme for orthogonal STBC[1] that uses a MIMO implementation. The scheme is summarized here for two transmitting antennas and one receiving antenna employing maximal ratio combining. The space–time view is illustrated in Figure 8.A.1 for an interval of time $(t, t + T)$. In this figure, symbol u_0 is transmitted over antenna 0 and symbol u_1 is simultaneously transmitted over transmitting antenna 1. In the next symbol period, $-u_1^*$ is sent from antenna 0 and u_0^* is sent from antenna 1.

The channel response between transmit antenna 0 and the receiver is denoted by $h_0(t)$, that is,

$$h_0(t) = h_0(t + T) = \alpha_0 e^{j\theta_0}$$

---

[1] S. Alamouti, A Simple Transmit Diversity Scheme for Wireless Communications, IEEE J Selected Areas in Comm, Oct. 1998, pp. 1451–1458.

	Space	
Time	Antenna 0	Antenna 1
t	u_0	u_1
$t+T$	$-u_1^*$	u_0^*

Figure 8.A.1 Space Time Block Coding.

where α_0 is the channel 0 amplitude and θ_0 is the channel 0 phase. Similarly, the channel response between transmit antenna 1 and the receiver is denoted by $h_1(t)$, that is,

$$h_1(t) = h_1(t+T) = \alpha_1 e^{j\theta_1}$$

where α_1 is the channel 1 amplitude and θ_1 is the channel 1 phase. The received signals from channels 0 and 1 are denoted by $r(t)$ and $r(t+T)$, respectively. Suppressing the time t and letting $r_0 = r(t)$ and $r_1 = r(t+T)$ allows the received signal at time t to be expressed as

$$r_0 = r(t) = h_0 u_0 + h_1 u_1 + z_0$$

and the signal at time $t+T$ to be

$$r_1 = r(t+T) = -h_0 u_1^* + h_1 u_0^* + z_1$$

where z_0 and z_1 are independent complex zero mean AWGN variables each with variance σ^2 added in channels 0 and 1, respectively. In matrix form, the aforementioned equations are written as

$$\begin{bmatrix} r_0 \\ r_1^* \end{bmatrix} = \begin{bmatrix} h_0 & h_1 \\ h_1^* & -h_0^* \end{bmatrix} \begin{bmatrix} u_0 \\ u_1 \end{bmatrix} + \begin{bmatrix} z_0 \\ z_1^* \end{bmatrix}$$

Letting $r = \begin{bmatrix} r_0 \\ r_1^* \end{bmatrix}$, $z = \begin{bmatrix} z_0 \\ z_1^* \end{bmatrix}$ and $H = \begin{bmatrix} h_0 & h_1 \\ h_1^* & -h_0^* \end{bmatrix}$, then $r = Hu + z$, where H is orthogonal.

Estimates of the transmitted symbols, denoted $\tilde{u} = \begin{bmatrix} \tilde{u}_0 \\ \tilde{u}_1 \end{bmatrix}$, are obtained by forming $\hat{u} = H^{T*} r$

Note that

$$H^{T*} H = \begin{bmatrix} h_0^* & h_1 \\ h_1^* & -h_0 \end{bmatrix} \begin{bmatrix} h_0 & h_1 \\ h_1^* & -h_0^* \end{bmatrix} = (|h_0|^2 + |h_1|^2) \begin{bmatrix} 1 & 0 \\ 0 & 1 \end{bmatrix}$$

$$= (\alpha_0^2 + \alpha_1^2) I_2 = D$$

SPACE–TIME BLOCK CODING FOR BPSK

where $I_2 = \begin{bmatrix} 1 & 0 \\ 0 & 1 \end{bmatrix}$ and the matrix $D = \begin{bmatrix} \alpha_0^2 + \alpha_1^2 & 0 \\ 0 & \alpha_0^2 + \alpha_1^2 \end{bmatrix}$ is a diagonal matrix. As a result,

$$\tilde{u} = H^{T*}r = H^{T*}Hu + H^{T*}z = Du + H^{T*}z$$

The average SNR $= (\alpha_0^2 + \alpha_1^2)\bar{\gamma}_b$. The estimates are then used to form the decision based on the minimum Euclidian distance.

For equal energy, BPSK signals choose u_0 if the squared Euclidian distance $d^2(\tilde{u}_0, u_0) \leq d^2(\tilde{u}_0, u_1)$, where $d^2(a, b) = |a|^2 + |b|^2 - 2Re(a^*b)$.

For the two transmit antennas each carrying one symbol and two symbols sent over successive intervals (*i.e.*, 2×2), the code rate $= 1$. In general with a block of n symbols transmitted (where n may be the number of transmit antennas) and m symbols in time, the code rate $R = n/m$.

Using Alamouti STBC, the BER performance can be estimated and compared to theoretical BPSK performance on a Rayleigh fading channel with diversity. For BPSK in Rayleigh fading with L independent diversity channels, the probability of bit error P_b is given by[2]

$$P_b = \left[\frac{1}{2}\left(1 - \sqrt{\frac{\bar{\gamma}_c}{1+\bar{\gamma}_c}}\right)\right]^L \sum_{k=0}^{L-1} \binom{L-1+k}{k} \left[\frac{1}{2}\left(1 + \sqrt{\frac{\bar{\gamma}_c}{1+\bar{\gamma}_c}}\right)\right]^k$$

where $\bar{\gamma}_b = L\bar{\gamma}_c$ and $\bar{\gamma}_c$ is the average SNR per channel. For large SNR $\bar{\gamma}_c$

$$P_b \approx \left(\frac{1}{4\bar{\gamma}_c}\right)^L \binom{2L-1}{L}$$

APPENDIX 8.B   SPACE–TIME BLOCK CODING FOR 16-QAM

BER performance for 16-QAM in Rayleigh fading has been investigated for two transmit antennas and L receive antennas.[3] The BER is expressed as

$$P_b = \frac{1}{2}(P_{b1} + P_{b3})$$

---

[2]Proakis, J.G. and M. Salehi, Digital Communications, 5th ed, McGraw-Hill, 2008, p. 854.
[3]Raju, M. S., A. Ramesh, and A. Chockalingam, BER Analysis of QAM with Transmit Diversity in Rayleigh Fading Channels, Proc of Globecom, 2003, pp. 641–645.

where $P_{b1} = \frac{1}{2}(P_1 + P_2)$, $P_{b3} = \frac{1}{2}(2P_1 + P_2 - P_3)$,

$$P_i = \left[\frac{1}{2}(1 - \mu_i)\right]^{2L} \sum_{k=0}^{2L-1} \binom{2L-1+k}{k} \left[\frac{1}{2}(1 + \mu_i)\right]^k, \ i = 1, 2, 3$$

and

$$\mu_1 = \sqrt{\frac{\overline{\gamma}_b}{5L + \overline{\gamma}_b}}, \ \mu_2 = \sqrt{\frac{9\overline{\gamma}_b}{5L + 9}} \text{ and } \mu_3 = \sqrt{\frac{25\overline{\gamma}_b}{5L + 25\overline{\gamma}_b}}$$

PROBLEMS

8.1 a. Find the theoretical BER for the parameters shown in Figure 8.1.

b. Using the high-SNR approximation find the theoretical BER for the parameters used in Figure 8.4 and compare with the simulated result.

c. Compute the theoretical result in item b above for $\bar{\gamma}_b = 20\,\text{dB}$ using the high-SNR assumption.

8.2 Using the relationships in Appendix 8.B.

Find the BER for $\bar{\gamma}_b = 10\,\text{dB}$ and $L = 1$.

Find the BER for $\bar{\gamma}_b = 10\,\text{dB}$ and $L = 2$.

8.3 Modify the Simulink model shown in Figure 8.7 to simulate BER results using maximum Doppler shift values = 0.1 and 0.001 Hz for the same range in $\bar{\gamma}_b$ as shown in Figure 8.8; include the theoretical results for diversity 4 and the maximum Doppler shift = 0.01 Hz

8.4 Modify the Simulink model in Figure 8.7 to use 2 transmit antennas and 1 receive antenna with the following model parameters:

Model Parameters for 16-QAM in Rayleigh Fading with STBC

- Sample time = symbol time = 1 s
- Frame based, 2 samples/frame
- Simulation time = stop with 100 errors
- Random integer seed = 22
- Jakes fading model with Doppler shift = 0.01 Hz
- Input signal power = 2 W
- Alamouti code: 2 trans, 1 rcv antennas (diversity = 2)

---

a. Display the Simulink model and the channel model.

b. Find the BER and compare the result with theoretical performance for an average SNR = 0 to 21 dB.

c. Repeat part b with Doppler shift = 0.001 Hz and find the BER.

9

DIGITAL COMMUNICATIONS BER PERFORMANCE IN AWGN (BLOCK CODING)

9.1 DIGITAL COMMUNICATIONS WITH BLOCK CODING IN AWGN

This chapter presents topics in the use of Simulink for evaluating block error control coding performance. Specific topics include:

- BER performance of BPSK in AWGN with the block error control codes
  - Hamming
  - Golay
  - Bose-Chaudhuri-Hocquenghem (BCH)
- BER performance of FSK in AWGN with Reed-Solomon (RS) codes
- BER performance of QAM in AWGN with Reed-Solomon codes
  - Multipath

9.2 BER PERFORMANCE OF BPSK IN AWGN WITH A BINARY BCH BLOCK CODE

A BCH code is a cyclic block code usually specified in terms of its generator polynomial. A binary BCH(n,k) code is represented by

---

Modeling of Digital Communication Systems Using SIMULINK®, First Edition.
Arthur A. Giordano and Allen H. Levesque.
© 2015 John Wiley & Sons, Inc. Published 2015 by John Wiley & Sons, Inc.
Companion Website: www.wiley.com/go/simulink

the following parameters:

$n = 2^m - 1 =$ code block length
$k =$ message length
$t =$ the number of correctable errors, $(n-k \leq mt)$
$d_{\min} \geq 2t + 1 =$ the minimum distance of the code
$R_c = k/n =$ code rate

For example, BCH(31,16) has a code rate 16/31 with a minimum distance $d_{\min} = 7$, capable of correcting $t = 3$ errors.

A Simulink model for estimating the BER performance of the BCH(31,16) code with BPSK modulation in AWGN is shown in Figure 9.1; for reference, a model for uncoded BPSK in AWGN is included in Figure 9.1. Model parameters are specified as follows:

Model Parameters for BCH(31,16) BPSK in AWGN

- BPSK antipodal signals $= +1$ and -1 $(M = 2)$
- BCH Symbol period $= 16/31$ s
- Sample based with sample time $= 1$ s
- Simulation time $= 10,000,000$ s
- Random integer seed $= 37$
- Input signal power $= 1$ W
- Computation delay $= 0$ s
- Receive delay $= 16$ s
- AWGN with $\gamma_b = 7$ dB, hard decisions $=>$
  - simulated BER for BCH BPSK $= 5.9 \times 10^{-5}$
  - simulated BER for uncoded BPSK $= 7.7 \times 10^{-4}$

Figure 9.2 displays the BCH encoder parameter selection where it is seen that the user can select a generator polynomial, a primitive polynomial, or a punctured code (used to shorten a code from its natural length). In Figure 9.1 defaults are assumed. Data from the random integer source is buffered into length 16 symbols for use by the BCH encoder. At the

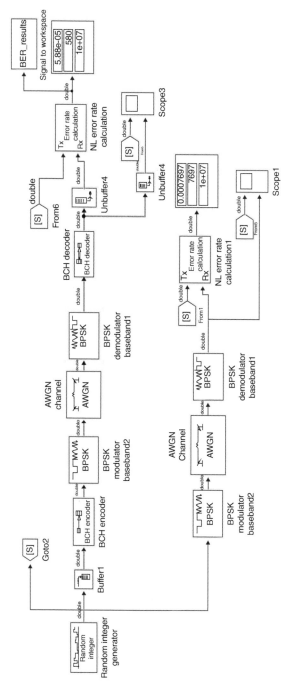

Figure 9.1 BER Estimation of BPSK with BCH(31,16) in AWGN.

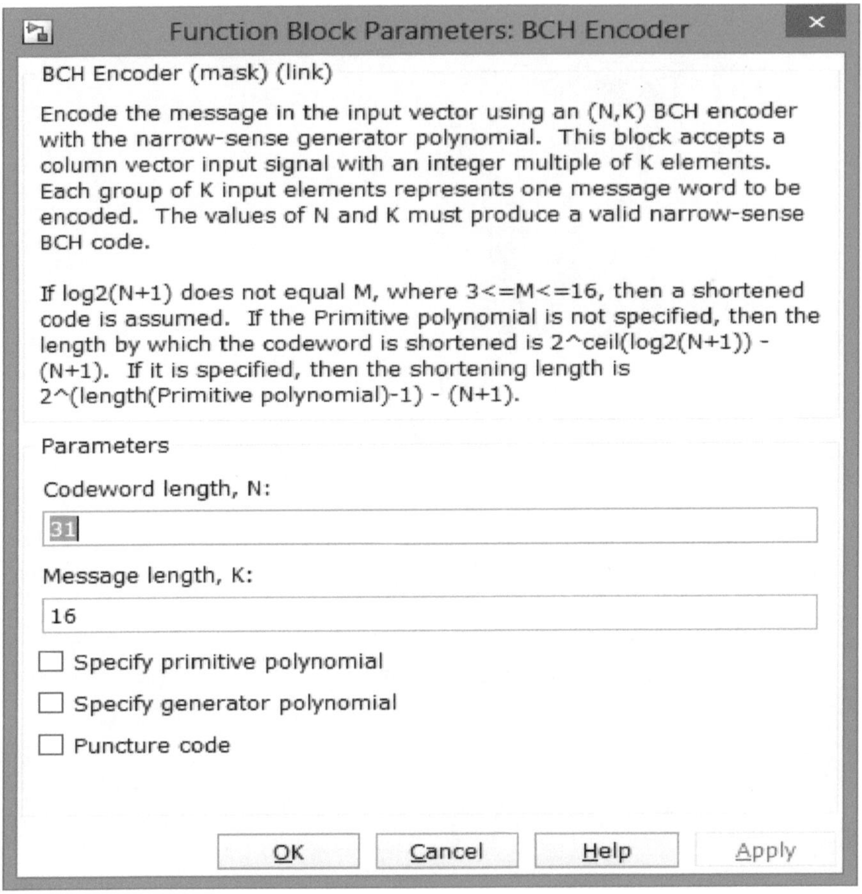

174 DIGITAL COMMUNICATIONS BER PERFORMANCE IN AWGN (BLOCK CODING)

Function Block Parameters: BCH Encoder

BCH Encoder (mask) (link)

Encode the message in the input vector using an (N,K) BCH encoder with the narrow-sense generator polynomial. This block accepts a column vector input signal with an integer multiple of K elements. Each group of K input elements represents one message word to be encoded. The values of N and K must produce a valid narrow-sense BCH code.

If log2(N+1) does not equal M, where 3<=M<=16, then a shortened code is assumed. If the Primitive polynomial is not specified, then the length by which the codeword is shortened is 2^ceil(log2(N+1)) - (N+1). If it is specified, then the shortening length is 2^(length(Primitive polynomial)-1) - (N+1).

Parameters

Codeword length, N:
31

Message length, K:
16

☐ Specify primitive polynomial
☐ Specify generator polynomial
☐ Puncture code

OK   Cancel   Help   Apply

Figure 9.2 BCH Encoder Parameters.

decoding end, the BCH decoder is then found to have a sequence output that is delayed by 16 s as shown in Figure 9.3 where Scope 3 displays the transmitted source data in the top trace and the decoder output in the bottom trace.

The selection of AWGN parameters for this simulation is shown in Figure 9.4. Here it is observed that the symbol signal to noise ratio $\gamma_s = \frac{16}{31}\gamma_b$ and the coded symbol period is 16/31 s.

The BER performance for BCH(31,16) with BPSK in AWGN using hard-decision decoding is shown in Figure 9.5 along with a theoretical upper bound on the coded BER performance. In general, the upper bound on the code word probability of error P_b for a t-error correcting, n-bit block code on a binary symmetric channel with transition probability p with hard decisions is given

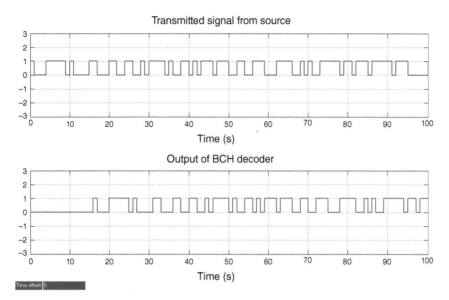

Figure 9.3 Output of Random Integer Source and Decoder Output.

by[1]

$$P_b \leq \frac{1}{n} \sum_{m=t+1}^{n} (m+t) \binom{n}{m} p^m (1-p)^{n-m}$$

where for BPSK $p = \frac{1}{2}\text{erfc}(\sqrt{R_c \gamma_b})$. For $\gamma_b = 7\,\text{dB}$, the aforementioned formula produces an upper bound (UB) BER $= 9.7 \times 10^{-5}$. Incorporating the BCH(31,16) code in conjunction with BPSK provides coding gain over the uncoded BPSK case that is most evident at high values of SNR.

9.3 BER PERFORMANCE OF BPSK IN AWGN WITH A HAMMING CODE

A Simulink model for estimating the BER performance for a Hamming(7,4) code with BPSK in AWGN is shown in Figure 9.6; for reference, a model for uncoded BPSK in AWGN is included in Figure 9.6.

The parameters for the Hamming encoder are shown in Figure 9.7. The Hamming(7,4) code is a binary, linear cyclic block code with $d_{\min} = 3$, $t = 1$,

---

[1] Michelson, A.M., and A H. Levesque, Error-Control Techniques for Digital Communications, John Wiley & Sons, 1985, p. 235.

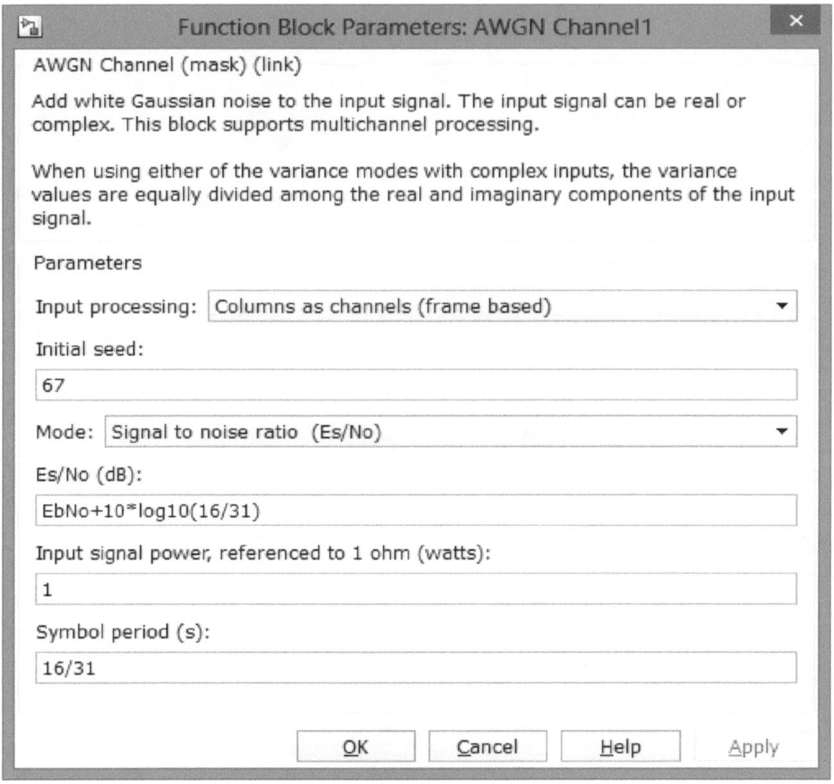

Figure 9.4 AWGN Parameter Selection.

and code rate $R_c = 4/7$. The default condition is selected to use a primitive polynomial over Galois field 2^m. (Note that m is capitalized in the Hamming encoder block parameters.)

Simulink model parameters are listed as follows

Model Parameters for Hamming(7,4) BPSK in AWGN

- BPSK antipodal signals $= +1$ and -1 ($M = 2$)
- Hamming encoder: m-degree primitive polynomial
- Hamming symbol period $= 4/7$ s
- Frame based with 4 samples/frame
- Sample time $= 1$ s
- Simulation time $= 1{,}000{,}000$ s

- Random integer seed = 37
- Input signal power = 1 W
- Computation delay = receive delay = 0 s
- AWGN with $\gamma_b = 7$ dB, hard decisions =>
  - simulated BER for Hamming(7,4)BPSK = 6.4×10^{-4}
  - simulated BER for uncoded BPSK = 7.7×10^{-4}

Figure 9.8 displays the BER performance for the Hamming(7,4) code with BPSK in AWGN.

The theoretical upper bound on the BER is obtained with the following formula

$$P_b \leq \frac{1}{7} \sum_{m=2}^{7} (m+1) \binom{7}{m} p^m (1-p)^{7-m}$$

where for BPSK $p = \frac{1}{2}\text{erfc}(\sqrt{R_c \gamma_b})$. For $\gamma_b = 7$ dB, the upper bound leads to a BER = 6.1×10^{-4}. For the Hamming(7,4) code with BPSK only slight coding gain is attained at high SNR over the uncoded case.

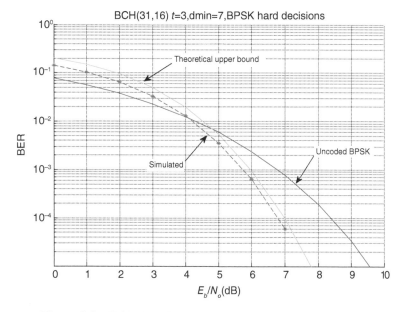

Figure 9.5 BCH(31,16) BPSK BER Performance in AWGN.

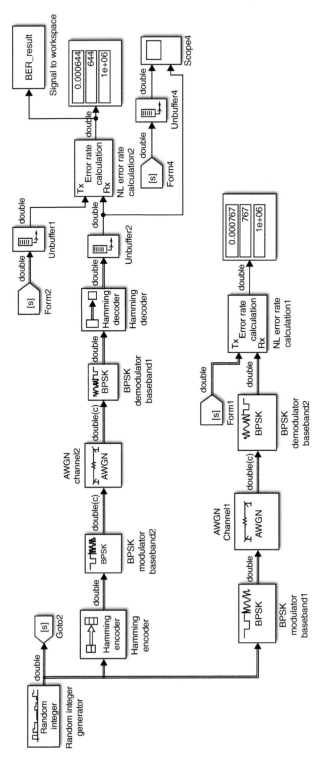

Figure 9.6 BER Estimation of Hamming (7,4) Code with BPSK in AWGN.

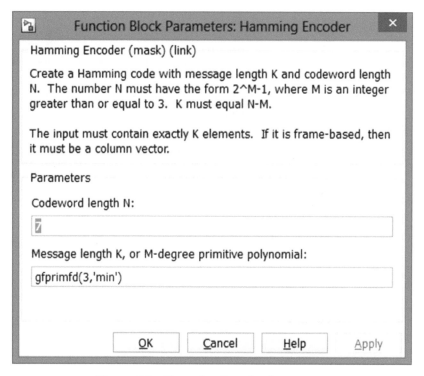

Figure 9.7 Hamming Encoder Parameters.

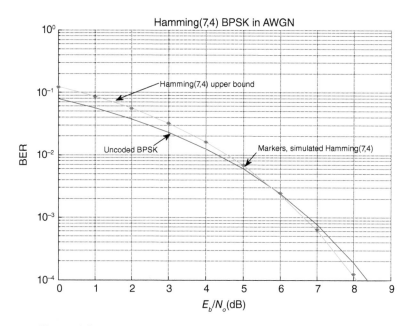

Figure 9.8 Hamming(7,4) BPSK BER Performance in AWGN.

9.4 BER PERFORMANCE OF BPSK IN AWGN WITH A GOLAY(24,12) BLOCK CODE

The extended Golay (24,12) code is a binary, linear cyclic block code with $d_{min} = 8$, $t = 3$, and code rate $R_c = 1/2$. A Simulink model for estimating the BER performance for Golay(24,12) code with BPSK in AWGN is shown in Figure 9.9; for reference uncoded BPSK in AWGN is included in Figure 9.9.

The Golay(24,12) encoder is implemented with a 12×24 generator matrix, where eye(12) is a 12×12 identity matrix and $G = [B \text{ eye}(12)]$, with B expressed as follows, that is[2],

```
B = [1 1 0 1 1 1 0 0 0 1 0 1
 1 0 1 1 1 0 0 0 1 0 1 1
 0 1 1 1 0 0 0 1 0 1 1 1
 1 1 1 0 0 0 1 0 1 1 0 1
 1 1 0 0 0 1 0 1 1 0 1 1
 1 0 0 0 1 0 1 1 0 1 1 1
 0 0 0 1 0 1 1 0 1 1 1 1
 0 0 1 0 1 1 0 1 1 1 0 1
 0 1 0 1 1 0 1 1 1 0 0 1
 1 0 1 1 0 1 1 1 0 0 0 1
 0 1 1 0 1 1 1 0 0 0 1 1
 1 1 1 1 1 1 1 1 1 1 1 0];
```

Note that care must be taken to separate the zeros and ones by spaces to allow Simulink to treat the entries as binary digits.

Simulink model parameters are specified as follows:

Model Parameters for Golay(24,12) BPSK in AWGN

- BPSK antipodal signals $= +1$ and -1 ($M = 2$)
- Golay encoder: $G = [B \text{ eye}(12)]$
- Golay symbol period $= 1/2$ s
- Frame based with 12 samples/frame
- Sample time $= 1$ s

---

[2]Michelson, A.M., and A H. Levesque, op. cit., p. 131.

- Simulation time = 1,000,000 s
- Random integer seed = 37
- Input signal power = 1 W
- Computation delay = receive delay = 0 s
- AWGN with $\gamma_b = 6$ dB, hard decisions =>
  - simulated BER for Golay(24,12) BPSK = 5.1×10^{-4}
  - simulated BER for uncoded BPSK = 2.4×10^{-3}

Figure 9.10 displays the BER performance for the Golay(24,12) code with BPSK in AWGN; coding gain is clearly observed in this figure. For $\gamma_b = 6$ dB, the theoretical upper bound BER = 6.1×10^{-4}.

9.5 BER PERFORMANCE OF FSK IN AWGN WITH REED-SOLOMON CODE

Reed-Solomon (RS) codes are nonbinary, linear cyclic codes where the code words are chosen from a q symbol alphabet. Letting $q = 2^m$, RS(n,k) code parameters are described by[3]

$$n = q - 1 = 2^m - 1, \, d_{min} = n - k + 1, \, t = \left\lfloor \frac{n-k}{2} \right\rfloor = \left\lfloor \frac{d_{min} - 1}{2} \right\rfloor$$

and a code rate $R_c = k/n$.

The specific RS code discussed here is RS(31,15) with $m = 5$, $t = 8$, $d_{min} = 17$, and $R_c = 15/31$. It is implemented in conjunction with 32-ary FSK orthogonal signals where each of the 2^5 symbols is mapped to one of the $M = 2^5$ orthogonal signals.

A Simulink model for estimating the BER performance for the RS(31,15) code with 32-FSK in AWGN is shown in Figure 9.11; for reference uncoded 32-FSK in AWGN is included in Figure 9.11. In the RS encoder and decoder, the default generator is selected and noncoherent detection with hard decisions is performed in the 32-FSK demodulator.

---

[3] $\lfloor x \rfloor$ denotes the largest integer contained in x

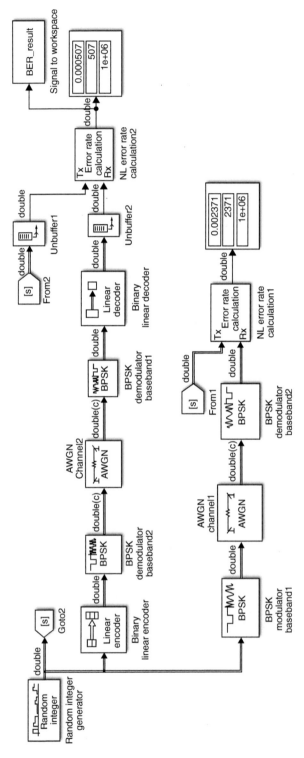

Figure 9.9 BER Estimation of Golay(24,12) Code with BPSK in AWGN.

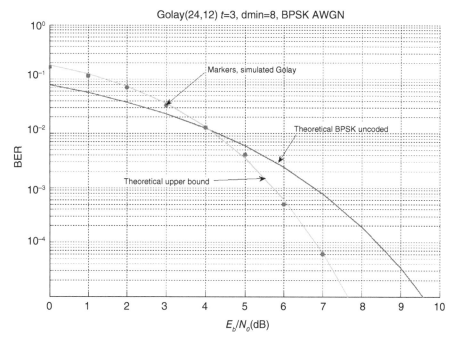

Figure 9.10 Golay(24,12) BPSK BER Performance in AWGN.

Simulink model parameters are listed as follows

Model Parameters for RS(31,15) 32-FSK in AWGN

- FSK orthogonal signals with $M = 32$
- RS symbol period $= 15/31$ s
- Frame based with 15 samples/frame
- Sample time $= 1$ s
- Simulation time $= 100{,}000$ s
- Random integer seed $= 37$
- Input signal power $= 1$ W
- $E_s/N_o = E_b/N_o + 10*\log((5 \times 15)/31)$
- Gain $K = 16/31$
- Computation delay $=$ receive delay $= 0$ s
- Hard decisions, noncoherent detection
- AWGN with $\gamma_b = 5$ dB, $\Rightarrow$
  - simulated BER for RS(31,15) 32-FSK $= 1.8 \times 10^{-3}$
  - simulated BER for uncoded 32-FSK $= 2.2 \times 10^{-3}$

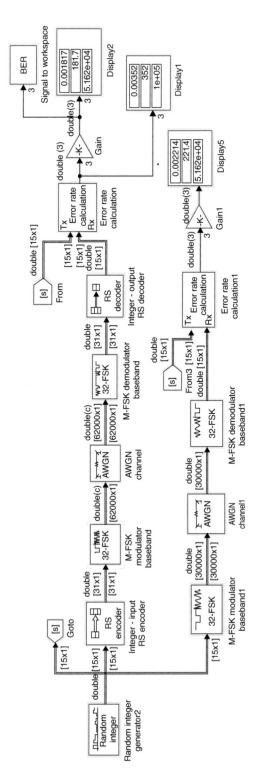

Figure 9.11 BER Estimation of RS(31,15) Code with 32-FSK in AWGN.

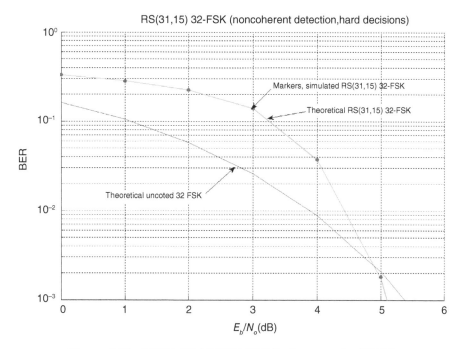

Figure 9.12 RS(31,15) 32-FSK BER Performance in AWGN.

Figure 9.12 displays the BER performance for the RS(31,15) code with 32-FSK in AWGN where it is observed that coding gain is realized at a BER level smaller than 2×10^{-3}.

The theoretical symbol error rate performance, P_{RS}, for the RS(31,15), 32-FSK in AWGN with noncoherent detection is given by[4]

$$P_{RS} = \frac{1}{n} \sum_{i=t+1}^{n} i \binom{n}{i} P_M^i (1-P_M)^{n-i}$$

where

$$P_M = \frac{1}{M} \sum_{l=2}^{M} (-1)^l \binom{M}{l} \exp\left[-\frac{(l-1) m R_c \gamma_b}{l}\right]$$

The corresponding BER is $P_b = \frac{2^{m-1}}{2^m - 1} P_{RS} = \frac{16}{31} P_{RS}$. Using $M = 32$, $m = 5$, $t = 8$, $R_c = 15/31$, $n = 31$, and $\gamma_b = 5$ dB, the theoretical BER $= 1.73 \times 10^{-3}$.

---

[4]Proakis, J.G., Digital Communications, 4th ed, McGraw-Hill, 2001, p. 465.

9.6 BER PERFORMANCE OF QAM IN AWGN WITH REED-SOLOMON CODING

This section addresses the BER performance of 16-QAM with RS(15,7) coding. This code has rate $R_c = 7/15$ with a minimum distance $d_{min} = 9$, capable of correcting up to $t = 4$ errors.

The Simulink model for estimating the BER for 16-QAM, RS(15,7) in AWGN is shown in Figure 9.13; for reference uncoded 16-QAM in AWGN is included in Figure 9.13.

Simulink model parameters are specified as follows:

Model Parameters for RS(15,7) 16-QAM in AWGN

- QAM signals with $M = 16$
- Sample time = 1 s
- Frame based with 7 samples/frame
- Simulation time = 1,000,000
- Symbol time = 7/15 s
- Random integer seed = 37
- Input signal power = 1 W
- $E_s/N_o = E_b/N_o + 10*\log(4 \times 7/15)$
- Gain $K = 1/4$
- Computation delay = receive delay = 0 s
- AWGN with $\gamma_b = 10$ dB =>
  - simulated BER for RS(15,7) 16-QAM = 4×10^{-4}
  - simulated BER for uncoded 16-QAM = 1.7×10^{-3}

Figure 9.14 displays the simulated and theoretical BER performance for RS(15,7) 16-QAM along with the theoretical uncoded 16-QAM performance.

The theoretical bit error rate performance, P_b, for the RS(15,7), 16-QAM in AWGN is given by[5]

$$P_b = \frac{1}{m} P_{RS} = \frac{1}{nm} \sum_{i=t+1}^{n} i \binom{n}{i} P_M^{\,i} (1 - P_M)^{n-i}$$

---

[5]Proakis, J. G., op. cit., p. 278 and p. 465.

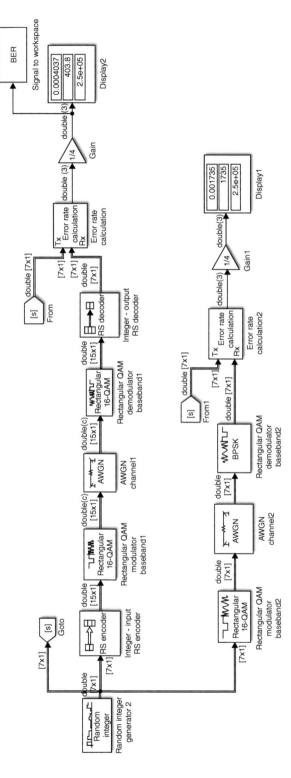

Figure 9.13 BER Estimation of RS(15,7) Code with 16-QAM in AWGN.

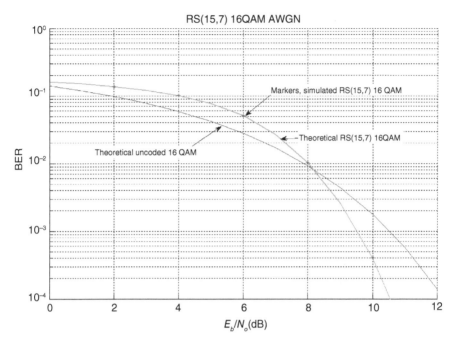

Figure 9.14 BER Performance for the RS(15,7) Code with 16-QAM in AWGN.

where

$$P_M = 2\left(1 - \frac{1}{\sqrt{M}}\right)\text{erfc}\left(\sqrt{\frac{3m\gamma_b R_c}{2(M-1)}}\right)$$

$$\left\{1 - \frac{1}{2}\left(1 - \frac{1}{\sqrt{M}}\right)\text{erfc}\left(\sqrt{\frac{3m\gamma_b R_c}{2(M-1)}}\right)\right\}$$

Using $M = 16$, $m = 4$, $t = 4$, $R_c = 7/15$, $n = 15$, and $\gamma_b = 10\,\text{dB}$, the theoretical BER $= 3.9 \times 10^{-4}$.

Another Simulink model for RS(15,7) with 16-QAM is shown in Figure 9.15 where a two-path multipath channel is included. The channel path coefficients are (1.0, 0.2) and the delay is 1 s. In Figure 9.15, the top line includes RS(15,7) coding with 16-QAM and two-path multipath, the second line includes 16-QAM and two-path multipath and the third line is uncoded 16-QAM with no multipath. The Simulink multipath channel is shown below the main Simulink model in Figure 9.15.

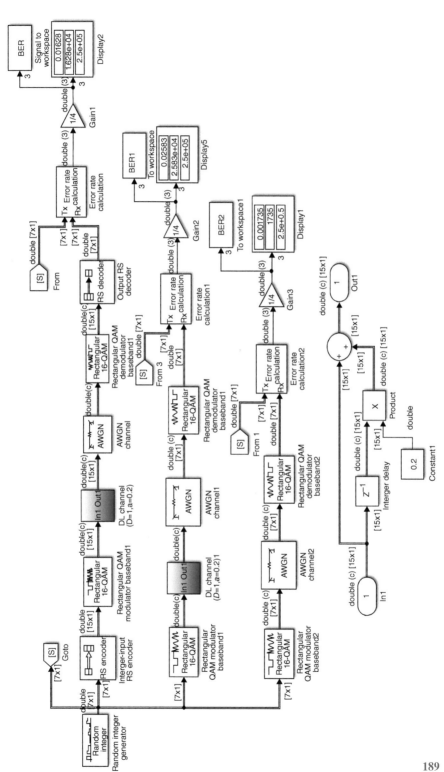

Figure 9.15 Simulink Model for 16-QAM with RS(15,7) and Multipath Channel(1,0,2).

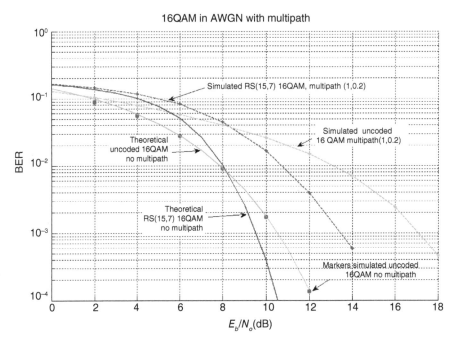

Figure 9.16 Impact from Multipath on BER for RS(15,7) 16-QAM.

The BER results are shown in Figure 9.16 for three simulated cases, that is, (i) RS(15,7)QAM with multipath, (ii) uncoded, 16-QAM with multipath, and (iii) uncoded 16-QAM with no multipath. For reference theoretical BERs are shown for RS(15,7) 16-QAM and uncoded16-QAM, both with no multipath. The results indicate that coding improves the BER when multipath is present but does not completely eliminate the degradation due to multipath.

9.7 SUMMARY DISCUSSION

This chapter has presented several examples of BER performance of block error control coding in conjunction with BPSK in AWGN. Both theoretical and simulated BER results were provided for specific binary block codes including BCH, Hamming, and Golay codes. The encoding/decoding was accomplished with either (i) default generator polynomial or (ii) user specified generator polynomial. Theoretical and simulated BER performance results were presented for nonbinary Reed-Solomon coding in conjunction with 32-FSK and 16-QAM. An example was provided to demonstrate the degradation due to multipath with and without coding. A summary of the simulated and theoretical BER results for each case is provided in Table 9.1.

TABLE 9.1 Simulated and Theoretical BER Results for Selected Codes

Code	d_{min}	t	Modulation	γ_b (dB)	Simulated	Theoretical
BCH(31,16)	7	3	BPSK	7	5.9×10^{-5}	9.7×10^{-5} (UB)
Hamming(7,4)	3	1	BPSK	7	6.4×10^{-4}	6.1×10^{-4} (UB)
Golay(24,12)	8	3	BPSK	6	5.1×10^{-4}	6.1×10^{-4} (UB)
RS(31,15)	17	8	32-FSK	5	1.8×10^{-3}	1.73×10^{-3}
RS(15,7)	9	4	16-QAM	10	4×10^{-4}	3.9×10^{-4}

PROBLEMS

9.1 a. Find the simulated BER for a BCH(15,11) code with BPSK in AWGN for $\gamma_b = 6$ dB.

b. Compare the simulated and theoretical result with uncoded BPSK in AWGN.

c. What is the minimum distance of the code?

d. What is the theoretical upper bound on the BER?

9.2 Select the binary linear encoder and decoder from the Communications System Toolbox and construct a Simulink model for the Hamming(7,4) code with BPSK. Use the Hamming code generator matrix [B eye(4)] with $B = [110;011;111;101]$. Use the same model parameters as those in the Figure 9.6 Simulink model. Find the BER for $\gamma_b = 7$ dB and compare the result to the value obtained in the Figure 9.6 Simulink model.

9.3 Sklar[6] has provided an approximate BER for a Golay(24,12) given by

$$P_b \approx \frac{1}{24} \sum_{m=4}^{24} m \binom{24}{m} p^m (1-p)^{24-m}$$

Compare the BER given by the aforementioned formula with the BER presented in Section 9.2 for $\gamma_b = 6$ dB.

9.4 Develop a Simulink model for RS(63,57) with 64-QAM in AWGN

a. Compare the simulated performance of the RS(63,57) 64-QAM case for uncoded 64-QAM with $\gamma_b = 14$ dB

b. For this code what is the minimum distance and number of correctable errors?

c. Estimate the theoretical post-decoding BER

---

[6] Sklar, B., Digital Communications, 2nd ed, Prentice Hall, 2001, p. 370.

10

DIGITAL COMMUNICATIONS BER PERFORMANCE IN AWGN (BLOCK CODING AND FADING)

10.1 DIGITAL COMMUNICATIONS WITH BLOCK CODING IN FADING

This chapter presents topics in Simulink based on block error control coding in a fading channel. Specific topics include the following:

- BER performance of BCH(31,16) BPSK in Rayleigh Fading with interleaving
- BER performance of Golay(24,12) BFSK in Rayleigh Fading with interleaving
- BER performance of Reed-Solomon(31,15) 32-FSK in Rayleigh Fading with interleaving
- BER performance of Reed-Solomon(15,7) 16-QAM in Rayleigh Fading with interleaving
- BER performance of Reed-Solomon(15,7) 16-QAM in Rician and Rayleigh Fading with interleaving
- BER performance in Rayleigh Fading with interleaving, block error control codes, and Alamouti STBC for

---

Modeling of Digital Communication Systems Using SIMULINK®, First Edition.
Arthur A. Giordano and Allen H. Levesque.
© 2015 John Wiley & Sons, Inc. Published 2015 by John Wiley & Sons, Inc.
Companion Website: www.wiley.com/go/simulink

- BCH(31,16) BPSK
- Golay(24,12) BFSK
- Reed-Solomon(31,15) 32-FSK
- Reed-Solomon(15,7) 16-QAM

10.2 BER PERFORMANCE OF BPSK IN RAYLEIGH FADING WITH INTERLEAVING AND A BCH BLOCK CODE

A Simulink model for estimating the BER for BPSK with a BCH(31,16) block code is shown in Figure 10.1a. The Rayleigh fading channel model is displayed in Figure 10.1b where a Jakes Doppler spectrum is selected. Figure 10.1c displays both the theoretical and simulated Doppler spectrum where the maximum Doppler shift is seen to be 0.1 Hz. The signal energy to noise power spectral density, E_s/N_o, is related to the bit energy to noise power spectral density, E_b/N_o, in terms of the code rate $R_c = 16/31$ according to $E_s/N_o = R_c E_b/N_o$. Note that the coded symbol period is 16/31 s for a 1 s sample time.

By executing the Simulink model with high $E_s/N_o = 100$ dB, the decoding delay in the BCH decoder can be determined. The scope is used to identify the misalignment between the transmitted and decoded sequences and thus obtain the delay. The input data sequence is delayed as seen in Figure 10.2 prior to performing the error rate computation.

The model parameters for the frame-based simulation are specified as follows:

Model Parameters for BCH(31,16) BPSK in Rayleigh Fading with Interleaving

- BPSK antipodal signals = +1 and −1 ($M = 2$)
- Symbol period = 16/31 s
- Sample time = 1 s
- Frame based with 16 samples/frame
- Interleaver 31 × 31
- Simulation time = stop with 200 errors
- Random integer seed = 22

- Input signal power = 1 W
- $E_s/N_o = E_b/N_o + 10\log(16/31)$
- Maximum Doppler shift = 0.1 Hz for Jakes fading
- Computation delay = Receive delay = 512 s

In order to compensate for the fading, an interleaver is required to produce independent (or near independent) bit errors. With 16 samples/frame the 31 × 31 interleaver spans 2 frames. Figure 10.3 illustrates the simulated BER for the BPSK BCH(31,16) case where a significant coding gain is seen over uncoded theoretical BPSK BER.

10.3 BER PERFORMANCE OF BFSK IN RAYLEIGH FADING WITH INTERLEAVING AND A GOLAY(24,12) BLOCK CODE

A Simulink model for estimating the BER for BFSK with a Golay(24,12) block code is shown in Figure 10.4. The Rayleigh Channel model and Jakes Doppler spectrum are the same as Figure 10.1b and c. The Golay(24,12) linear encoder generator matrix is expressed as [B eye(12)] where

```
B = [1 1 0 1 1 1 0 0 0 1 0 1
1 0 1 1 1 0 0 0 1 0 1 1
0 1 1 1 0 0 0 1 0 1 1 1
1 1 1 0 0 0 1 0 1 1 0 1
1 1 0 0 0 1 0 1 1 0 1 1
1 0 0 0 1 0 1 1 0 1 1 1
0 0 0 1 0 1 1 0 1 1 1 1
0 0 1 0 1 1 0 1 1 1 0 1
0 1 0 1 1 0 1 1 1 0 0 1
1 0 1 1 0 1 1 1 0 0 0 1
0 1 1 0 1 1 1 0 0 0 1 1
1 1 1 1 1 1 1 1 1 1 1 0];
```

The sample time and symbol period are 1 and $1/2$ s, respectively. A 24 × 24 interleaver is incorporated in the simulation and the FSK tones are spaced at 10 Hz.

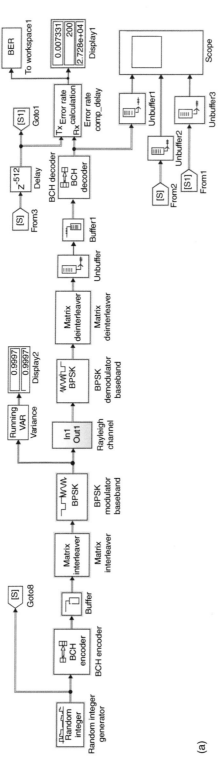

Figure 10.1 (a) Simulink Model for BCH(31,16) BPSK BER in Rayleigh Fading with Interleaving ($E_b/N_o = 10\,\text{dB} = >$ Simulated BER = 0.007); (b) Rayleigh Channel in Simulink Model for BCH(31,16) BPSK BER in Rayleigh Fading with Interleaving; (c) Jakes Doppler Spectrum for Rayleigh Fading Channel.

BER PERFORMANCE OF BFSK IN RAYLEIGH FADING

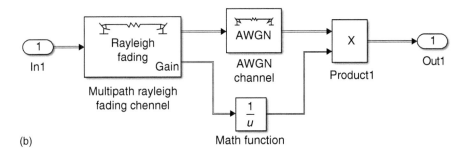

(b)

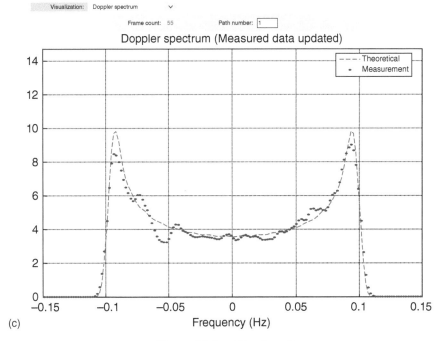

(c)

Figure 10.1 (*Continued*)

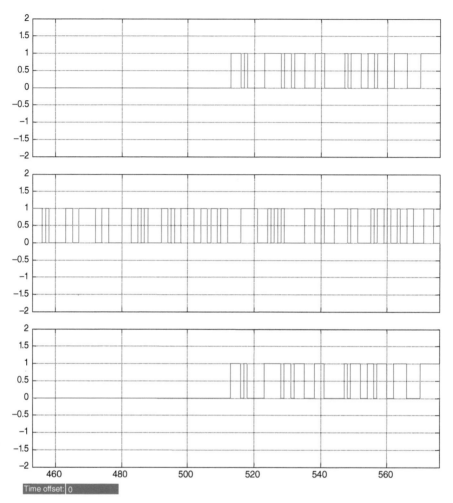

Figure 10.2 Scope Used to Set Receive Delay with $E_b/N_o = 100\,\text{dB}$.[1]

The model parameters are specified as follows:

Model Parameters for Golay(24,12) BFSK in Rayleigh Fading

- BFSK 10 Hz tone spacing
- Symbol period = 1/2 s
- Sample time = 1 s

[1]The scope, used to set the receive delay, might not be shown in subsequent models to simplify the model configuration.

- Frame based with 12 samples/frame
- Linear encoder and decoder [B eye(12)]
- Interleaver 24 × 24
- Simulation time = stop with 200 errors
- Random integer seed = 22
- Input signal power = 1 W
- $E_s/N_o = E_b/N_o + 10\log(1/2)$
- Maximum Doppler shift = 0.1 Hz for Jakes fading
- Computation delay = Receive delay = 300 s

Odenwalder has developed a theoretical formulation for the BER using a Golay(24,12) code. The probability of bit error P_b in terms of channel error probability p with hard decisions is given by[2]

$$P_b = \frac{1}{24} \sum_{i=4}^{24} \beta_i \binom{24}{i} p^i (1-p)^{24-i}$$

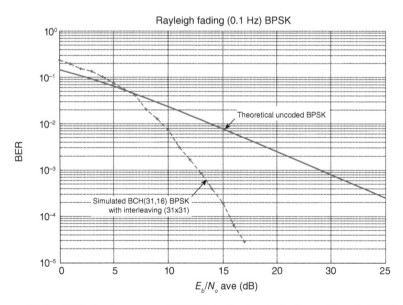

Figure 10.3 BER Performance of BPSK in Rayleigh Fading with a BCH(31,16) Block Code.

[2]Odenwalder, J.P. "Error Control Coding Handbook," Final Report, Contract No.F44620-76-0056, Linkabit Corporation, LaJolla, CA July 1976.

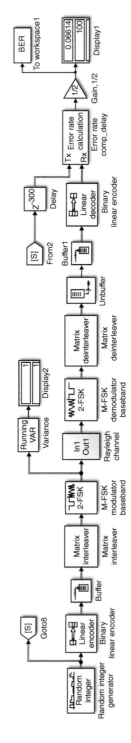

Figure 10.4 Simulink Model for Golay(24,12) BFSK BER in Rayleigh Fading with Interleaving ($E_b/N_o = 10\,\text{dB} \Rightarrow$ Simulated BER = 0.066).

where the vector of coefficients

$$\beta_i = \begin{bmatrix} 0 & 0 & 4 & 8 & {}^{120}/_{19} & 8 & 8 & {}^{2637}/_{323} & {}^{3256}/_{323} & {}^{3656}/_{323} & 12 & {}^{4096}/_{323} & {}^{4496}/_{323} & {}^{5115}/_{323} \\ 16 & 16 & {}^{336}/_{19} & 16 & 20 & 24 & 24 \end{bmatrix}$$

In Rayleigh Fading, assuming Golay(24,12) coding and BFSK with non-coherent detection, the channel error probability p is given by

$$p = \frac{1}{2 + R_c \overline{\gamma_b}}$$

where the code rate $R_c = \frac{12}{24}$ and the average E_b/N_o is $\overline{\gamma_b} = \frac{E_b}{N_o}$

Another estimate for the BER has been identified by Sklar.[3] Sklar indicates that a good approximation for the probability of bit error P_b is given by

$$P_b = \frac{1}{24} \sum_{i=4}^{24} i \binom{24}{i} p^i (1-p)^{24-i}$$

Figure 10.5 depicts the simulated BER for BFSK with the Golay(24,12) code in Rayleigh fading assuming the 24 × 24 interleaver. This shows good agreement with Odenwalder's theoretical BER results.

To investigate the effects of using a longer interleaver, replace the 24 × 24 interleaver with a 48 × 48 interleaver. The BER results for both interleavers are shown in Figure 10.6 where it is observed that the interleaver choice has little impact on the BER results for the Rayleigh fading model.

It is also instructive to compare the BFSK Golay(24,12) BER results with uncoded BFSK results implemented with a diversity scheme. Figure 10.7 displays the simulated results along with uncoded BFSK having diversity order of 1, 2, and 4. The figure indicates that the BFSK Golay(24,12) case is similar to uncoded BFSK with dual diversity.

10.4 BER PERFORMANCE OF 32-FSK IN RAYLEIGH FADING WITH INTERLEAVING AND A REED-SOLOMON(31,15) BLOCK CODE

In this section, a higher order modulation is combined with a nonbinary Reed-Solomon (RS) code incorporated in the Simulink model displayed

---

[3] Sklar, B., Digital Communications, 2nd ed, Prentice Hall PTR, Uper Saddle River, NJ, p. 370.

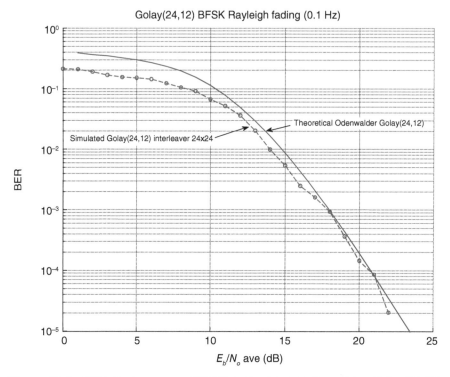

Figure 10.5 BER Performance of BFSK in Rayleigh Fading with a Golay(24,12) Block Code and 24 × 24 Interleaver.

in Figure 10.8. With a 32-FSK modulation, an appropriate match is an RS(31,15) code, which has a minimum distance of 17. Since there are 2^5 possible FSK symbols in the 32-ary alphabet, $E_s/N_o = (5)(15/31)E_b/N_o$. The Rayleigh channel model is the same as that in Figure 10.1b and c and a 31 × 31 interleaver is selected. The sample and symbol period are 1 s and 15/31 s, respectively.

The model parameters are specified as follows:

Model Parameters for RS(31,15) 32-FSK in Rayleigh Fading

- Frequency separation = 100 Hz
- 1000 samples/symbol
- Sample time = 1 s
- Symbol period = 15/31 s
- Frame based with 15 samples/frame

- Simulation time = stop with 100 errors
- Input signal power = 1 W
- $E_s/N_o = E_b/N_o + 10*\log(5*15/31)$
- Interleaver = 31×31
- Gain, $K = 16/31$
- Maximum Doppler shift = 0.1 Hz for Jakes fading
- Computation delay = receive delay = 480 s

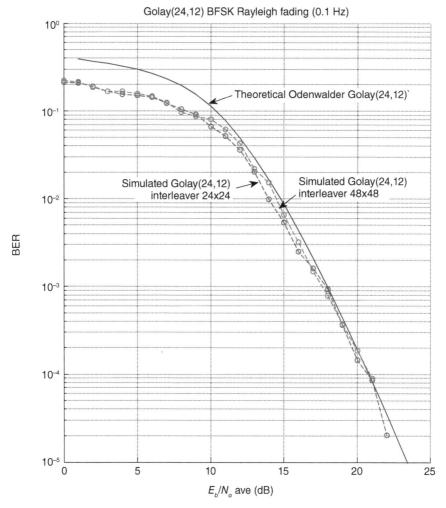

Figure 10.6 BER Performance of BFSK in Rayleigh Fading with a Golay(24,12) Block Code using 24×24 and 48×48 Interleavers.

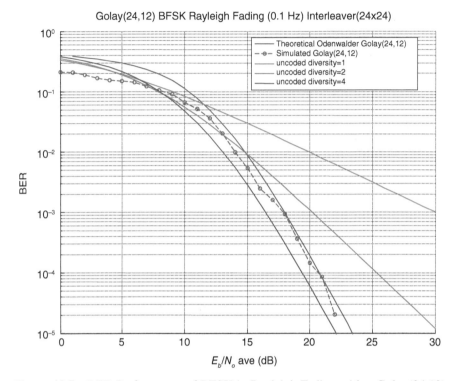

Figure 10.7 BER Performance of BFSK in Rayleigh Fading with a Golay(24,12) Block Code with 24 × 24 Interleaver and Uncoded BFSK with Diversity.

Figure 10.9 compares the simulated BER performance for the 32-FSK RS(31,15) case along with the theoretical uncoded FSK performance. Substantial coding gain is evident from this figure.

10.5 BER PERFORMANCE OF 16-QAM IN RAYLEIGH FADING WITH INTERLEAVING AND A REED-SOLOMON(15,7) BLOCK CODE

In this section, the Simulink model displayed in Figure 10.10 is constructed with a 16-QAM modulation matched to a RS(15,7) code (minimum distance 9) with an associated 30 × 30 interleaver. Then, $E_s/N_o = (4)(7/15)E_b/N_o$. The sample and symbol period are 1 s and 7/15 s, respectively. The Rayleigh channel model is again the one used in Figure 10.1b and c.

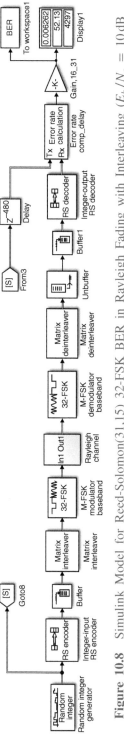

Figure 10.8 Simulink Model for Reed-Solomon(31,15) 32-FSK BER in Rayleigh Fading with Interleaving ($E_b/N_o = 10\,\text{dB}$ => Simulated BER = 0.0063).

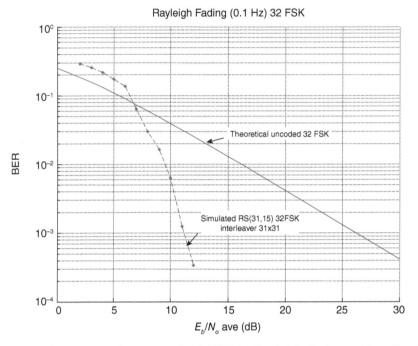

Figure 10.9 BER Performance of 32-FSK in Rayleigh Fading with a Reed-Solomon(31,15) Block Code and 31 × 31 Interleaver.

The model parameters are specified as follows:

Model Parameters for RS(15,7) 16-QAM in Rayleigh Fading

- Symbol period = 7/15 s
- Sample time = 1 s
- Frame based with 7 samples/frame
- Simulation time stop with 200 errors
- Input signal power = 1 W
- $E_s/N_o = E_b/N_o + 10\log(4) + 10\log(7/15)$
- Interleaver 30 × 30
- Maximum Doppler shift = 0.1 Hz for Jakes fading
- Computation delay = Receive delay = 427 s

The simulated 16-QAM RS(15,7) BER performance is compared with theoretical uncoded BER performance in Figure 10.11. Once again, substantial coding gain is observed in the figure.

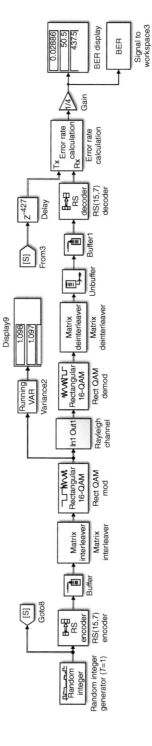

Figure 10.10 Simulink Model for Reed-Solomon(15,7) 16-QAM BER in Rayleigh Fading with Interleaving ($E_b/N_o = 10\,\text{dB}$ => Simulated BER = 0.029).

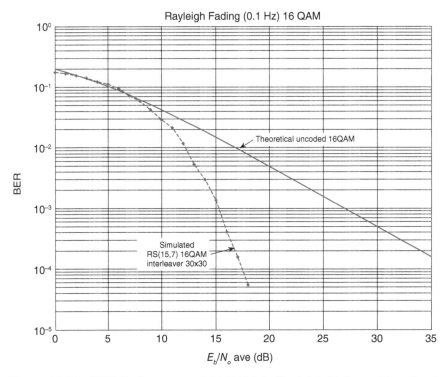

Figure 10.11 BER Performance of 16-QAM in Rayleigh Fading with a Reed-Solomon(15,7) Block Code and 30 × 30 Interleaver.

A second channel model is now considered, where the Rayleigh fading channel is changed to introduce a Jakes Doppler spectrum with a 0.01 Hz maximum Doppler shift. The BER results are shown in Figure 10.12 where it is seen that the fading occurs over a longer interval so that the interleaver is not compensating as effectively for the fades; consequently, a poorer BER is observed in the 0.01 Hz case.

10.6 BER PERFORMANCE OF 16-QAM IN RAYLEIGH AND RICIAN FADING WITH INTERLEAVING AND A REED-SOLOMON(15,7) BLOCK CODE

In many instances, Rician rather than Rayleigh fading occurs in the channel.[4] BER results are now obtained using the Simulink model in

---

[4]Rician fading is characterized by a K factor where ordinarily a strong dominant component is present.

BER PERFORMANCE OF 16-QAM IN RAYLEIGH AND RICIAN FADING

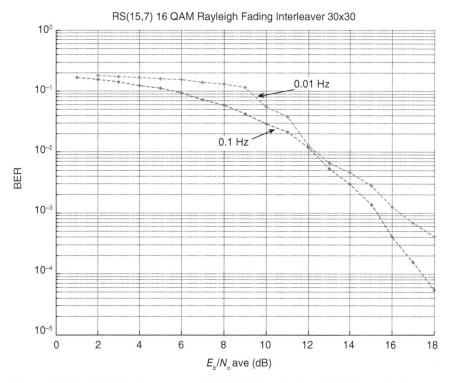

Figure 10.12 BER Performance of 16-QAM in Rayleigh Fading with 0.1 and 0.01 Hz and with a Reed-Solomon(15,7) Block Code and 30 × 30 Interleaver.

Figure 10.13 for 16QAM RS(15,7) and a 30 × 30 interleaver. Two separate paths are implemented where the only difference is the choice of a Rician or Rayleigh channel. The Rician channel is modeled with a Jakes fading spectrum having the maximum diffuse Doppler shift and the line-of-sight Doppler shift both selected to be 0.1 Hz.

The model parameters are specified as follows:

Model Parameters for RS(15,7) 16 QAM in Rayleigh and Rician Fading

- Symbol period = 7/15 s
- Sample time = 1 s
- Frame based with 7 samples/frame
- Simulation time stop = 100 errors (Rice)

- Input signal power = 1 W
- $E_s/N_o = E_b/N_o + 10\log(4) + 10\log(7/15)$
- Interleaver 30 × 30
- Rayleigh Maximum Doppler shift = 0.1 Hz for Jakes fading
- Rice (Jakes fading)
  - max diffuse Doppler shift = 0.1 Hz
  - LOS Doppler shift = 0.1 Hz
  - Rice K factor = 1 and 3
- Computation delay = receive delay = 427 s

The results comparing Rayleigh fading along with Rician fading using Rice K factors of 1 and 3 are shown in Figure 10.14. In this figure, it is observed that Rayleigh fading produces the poorest BER results.

10.7 BER PERFORMANCE OF BPSK IN RAYLEIGH FADING WITH INTERLEAVING AND A BCH BLOCK CODE AND ALAMOUTI STBC

As discussed in Chapter 8, diversity techniques can improve performance when the diversity branches exhibit independent fading. Multiple antennas for the transmitter, the receiver, or both offer this improvement without expanding the bandwidth and allow for an increase in data rate. As introduced in Chapter 8, Alamouti STBC is utilized. In a two-antenna transmit diversity scheme with a single receive antenna, two symbols are sent simultaneously over the two transmit antennas and are then resent after space–time encoding in the next symbol interval; the receiver then combines the symbols over two symbol intervals and makes a decision using a maximum likelihood detector.

The Simulink model shown in Figure 10.15a uses a 2 × 2 Alamouti STBC scheme that surrounds the 2 × 2 Rayleigh channel model displayed in detail in Figure 10.15b. The four Rayleigh channel paths all assume a Jakes Doppler spectrum with a maximum Doppler shift of 0.1 Hz. The transmit waveform is a BCH(31,16) BPSK scheme that includes a 31 × 31 interleaver. The sample time and symbol period are 1 s and 16/31 s, respectively. Since the Alamouti combiner receives two equal-power symbols, the E_s/N_o is adjusted by a factor of two for estimating the BER.

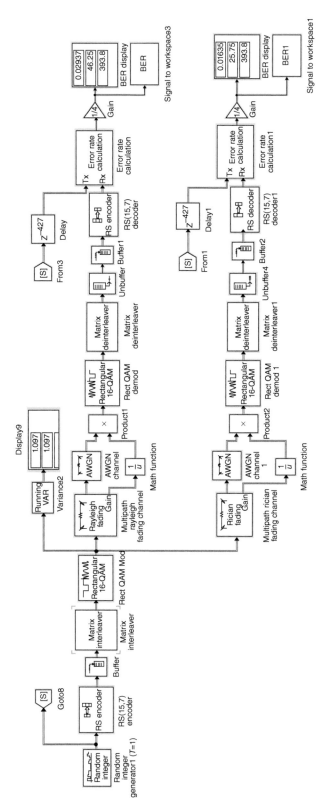

Figure 10.13 Simulink Model for Reed-Solomon(15,7) 16-QAM BER in Rayleigh and Rician Fading with Interleaving (Rayleigh Fading: $E_b/N_o = 10\,\text{dB} \Rightarrow$ Simulated BER(185 Errors) = 0.029; Rician Fading: $E_b/N_o = 10\,\text{dB} \Rightarrow$ Simulated BER(100 Errors) = 0.016).

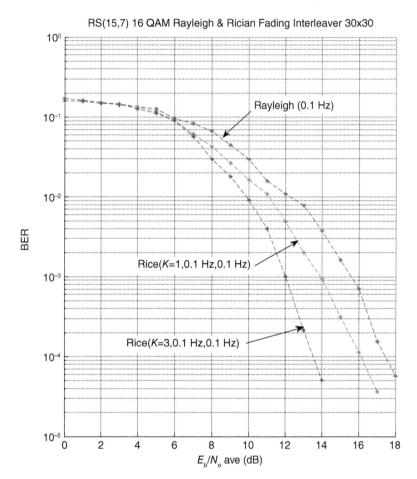

Figure 10.14 BER Performance of 16-QAM in Rayleigh Fading with 0.1 Hz and Rician Fading with Maximum Diffuse Doppler Shift = 0.01 Hz and Line-of-Sight Doppler Shift = 0.1 Hz with a Reed-Solomon(15,7) Block Code and 30 × 30 Interleaver.

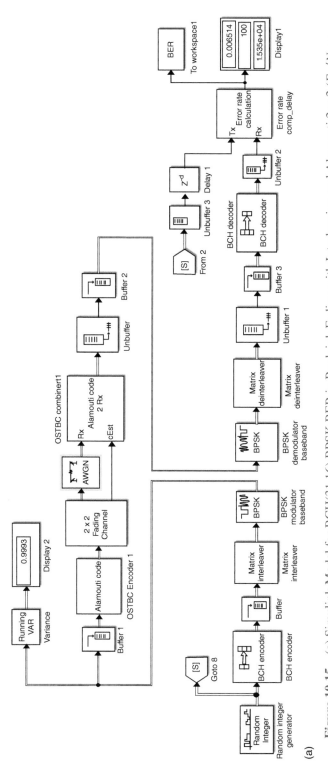

Figure 10.15 (a) Simulink Model for BCH(31,16) BPSK BER in Rayleigh Fading with Interleaving and Alamouti 2×2 ($E_b/N_o = 6\,\text{dB} \Longrightarrow$ Simulated BER = 0.0065); (b) Expanded View of 2×2 Rayleigh Fading Channel.

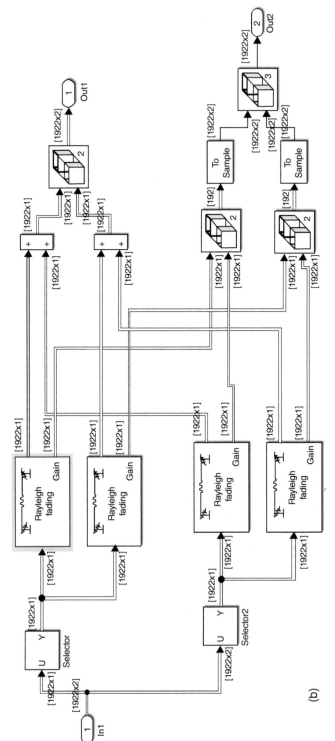

Figure 10.15 (*Continued*)

The model parameters are specified as follows:

Model Parameters for BCH(31,16) BPSK in Rayleigh Fading with STBC

- BPSK antipodal signals = +1 and −1 ($M = 2$)
- Symbol period = 16/31 s
- Sample time = 1 s
- Frame based with 16 samples/frame
- Simulation time = stop with 100 errors
- Alamouti 2×2
- Input signal power = 1 and 2 W into combiner
- $E_s/N_o = E_b/N_o - 10\log(2) + 10\log(16/31)$
- Maximum Doppler shift = 0.1 Hz in 4 paths (for Jakes fading)
- Interleaver 31×31
- Computation delay = receive delay = 2000 s

The BER performance for the BCH(31,16) BPSK with 31×31 interleaving and 2×2 Alamouti STBC is shown in Figure 10.16 along with the theoretical uncoded BPSK BER assuming diversity 4. The performance gain from the use of STBC over the uncoded diversity 4 case is apparent for good SNR with a small loss in performance at poor SNR.

10.8 BER PERFORMANCE OF BFSK IN RAYLEIGH FADING WITH INTERLEAVING AND A GOLAY(24,12) BLOCK CODE AND ALAMOUTI STBC

The Simulink model shown in Figure 10.17 implements 2×2 Alamouti STBC for the Golay(24,12) code with BFSK and a 24×24 interleaver. The four-path Rayleigh channel model is the same as the one used in Figure 10.15b.

The model parameters are specified as follows:

Model Parameters for Golay(24,12) BFSK in Rayleigh Fading with STBC

- BFSK 10 Hz tone spacing
- Symbol period = 1/2 s

- Sample time = 1 s
- Frame based with 12 samples/frame
- Linear encoder and decoder [B eye(12)]
- Simulation time = stop with 100 errors
- Alamouti 2×2
- Input signal power = 1 and 2 W into combiner
- $E_s/N_o = E_b/N_o - 10\log(2) + 10\log(1/2)$
- Maximum Doppler shift = 0.1 Hz in 4 paths (Jakes fading)
- Interleaver 24×24
- Computation delay = receive delay = 300 s

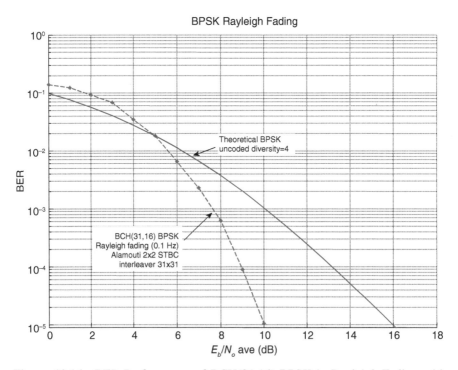

Figure 10.16 BER Performance of BCH(31,16) BPSK in Rayleigh Fading with Alamouti 2×2 and 31×31 Interleaving.

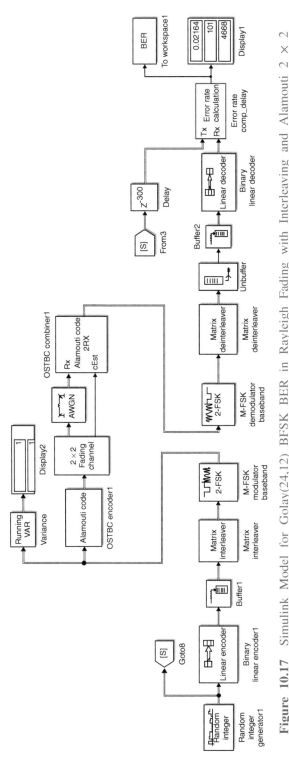

Figure 10.17 Simulink Model for Golay(24,12) BFSK BER in Rayleigh Fading with Interleaving and Alamouti 2×2 ($E_b/N_o = 10\,\text{dB} =>$ Simulated BER $= 0.022$).

As presented in Section 10.3, the Golay(24,12) linear encoder generator matrix is expressed as [B eye(12)] where

```
B = [1 1 0 1 1 1 0 0 0 1 0 1
1 0 1 1 1 0 0 0 1 0 1 1
0 1 1 1 0 0 0 1 0 1 1 1
1 1 1 0 0 0 1 0 1 1 0 1
1 1 0 0 0 1 0 1 1 0 1 1
1 0 0 0 1 0 1 1 0 1 1 1
0 0 0 1 0 1 1 0 1 1 1 1
0 0 1 0 1 1 0 1 1 1 0 1
0 1 0 1 1 0 1 1 1 0 0 1
1 0 1 1 0 1 1 1 0 0 0 1
0 1 1 0 1 1 1 0 0 0 1 1
1 1 1 1 1 1 1 1 1 1 1 0];
```

The simulated BER for the Golay(24,12) BFSK with a 24 × 24 interleaver and 2 × 2 Alamouti STBC is shown in Figure 10.18 along with the BER for uncoded BFSK with diversity 4. The performance benefit of using STBC is easily seen.

10.9 BER PERFORMANCE OF 32-FSK IN RAYLEIGH FADING WITH INTERLEAVING AND A REED-SOLOMON(31,15) BLOCK CODE AND ALAMOUTI STBC

The Simulink model shown in Figure 10.19 implements 2 × 2 Alamouti STBC for the Reed-Solomon (31,15) code with 32-FSK and a 31 × 31 interleaver. The four-path Rayleigh channel model is the same as the one shown in Figure 10.15b.

The model parameters are specified as follows:

Model Parameters for RS(31,15) 32-FSK in Rayleigh Fading with STBC

- Frequency separation = 100 Hz
- 1000 samples/symbol
- Sample time = 1 s
- Symbol period = 15/31 s
- Frame based with 15 samples/frame
- Simulation time stop = 100 errors

- STBC 2 × 2 Alamouti
- Input signal power = 1 and 2 W into combiner
- $E_s/N_o = E_b/N_o - 10\log(2) + 10*\log(5*15/31)$
- Maximum Doppler shift = 0.1 Hz in 4 paths (Jakes fading)
- Interleaver 31 × 31
- Computation delay = receive delay = 480 s

The simulated BER for the RS(31,15) 32-FSK with a 31 × 31 interleaver and 2 × 2 Alamouti STBC is shown in Figure 10.20 along with the BER for uncoded 32-FSK with diversity 4. The performance benefit of using STBC is again observed.

10.10 BER PERFORMANCE OF 16-QAM IN RAYLEIGH FADING WITH INTERLEAVING AND A REED-SOLOMON (15,7) BLOCK CODE AND ALAMOUTI STBC

The Simulink model shown in Figure 10.21 implements 2 × 2 Alamouti STBC for the RS(15,7) code with 16-QAM and a 30 × 30 interleaver. The four path Rayleigh channel model is the same as the one shown in Figure 10.15b.

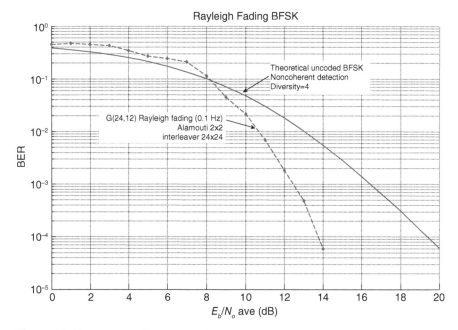

Figure 10.18 BER Performance of Golay(24,12) BFSK in Rayleigh Fading with Alamouti 2 × 2 and Interleaver 24 × 24.

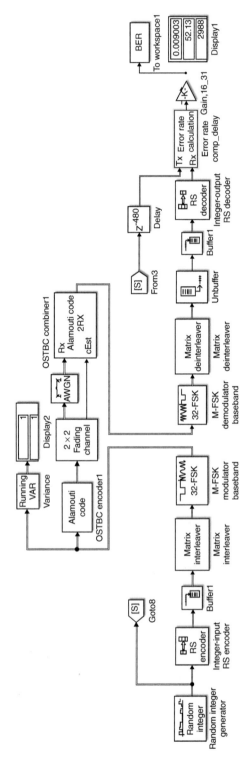

Figure 10.19 Simulink Model for RS(31,15) 32-FSK BER in Rayleigh Fading with Interleaving and Alamouti 2×2 STBC ($E_b/N_o = 6\,\text{dB} = >$ Simulated BER = 0.009).

BER PERFORMANCE OF 16-QAM IN RAYLEIGH FADING   221

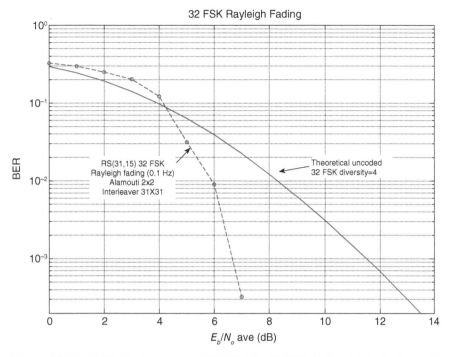

Figure 10.20 BER Performance of RS(31,15) 32-FSK in Rayleigh Fading with Alamouti 2×2 and Interleaver 31×31.

The model parameters are specified as follows:

Model Parameters for RS(15,7) 16 QAM in Rayleigh Fading with STBC

- Symbol period = 7/15 s
- Sample time = 1 s
- Frame based with 7 samples/frame
- Simulation time stop = 100 errors
- STBC 2×2 Alamouti
- Input signal power = 1 and 2 W into combiner
- $E_s/N_o = E_b/N_o - 10\log(2) + 10\log(4*7/15)$
- Maximum Doppler shift = 0.1 Hz in 4 paths (Jakes fading)
- Interleaver 30×30
- Computation delay = receive delay = 427 s

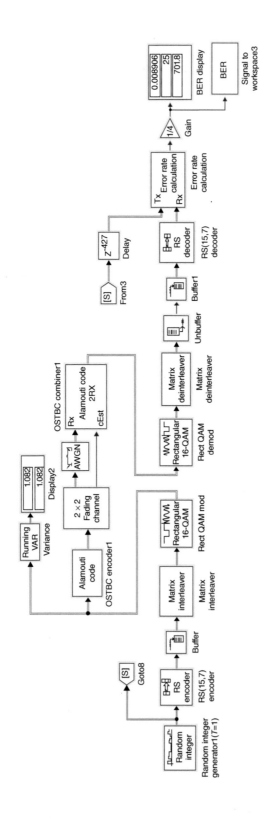

Figure 10.21 Simulink Model for RS(15,7) 16-QAM BER in Rayleigh Fading with Interleaving and Alamouti 2×2 STBC ($E_b/N_o = 10\,\text{dB} => $ Simulated BER $= 0.009$).

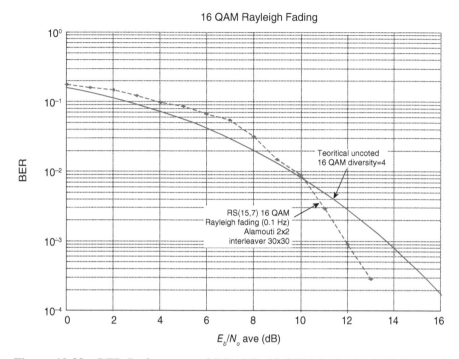

Figure 10.22 BER Performance of RS(15,7) 16-QAM in Rayleigh Fading with Alamouti 2 × 2 STBC and Interleaver 30 × 30.

The simulated BER for RS(15,7) 16-QAM with a 30 × 30 interleaver and 2 × 2 Alamouti STBC is shown in Figure 10.22 along with the BER for uncoded 16QAM with diversity 4. The performance benefit of using STBC is observed again.

10.11 SUMMARY DISCUSSION

This chapter has presented BER results for various combinations of block codes and modulations transmitted over fading channels. By incorporating interleaving, the block error control codes can utilize their individual distance properties to correct bit errors and provide significant coding gain over uncoded schemes that use the same modulation. Most of the results were developed for Rayleigh fading but an example with Rician fading was included, to demonstrate that Rayleigh fading causes the worst case BER.

The last sections of this chapter repeated the block error control coding and modulation cases with interleaving and Alamouti STBC. The STBC results indicate significant gain over uncoded modulations with large diversity order.

PROBLEMS

10.1 This problem is constructed to estimate BER results with and without interleaving. Develop a Simulink model with Rayleigh fading for BCH(31,16) BPSK with a maximum Doppler = 0.1 Hz

   a. Identify the model parameters and display the Simulink model

   b. Simulate the BER with no interleaving using steps of 1 dB between 0 and 21 dB

   c. Simulate the BER with a 31 × 31 interleaver using steps of 1 dB between 0 and 17 dB

   d. Compare the results with theoretical uncoded BPSK

10.2 Develop and display a Simulink model with Rayleigh fading for Hamming(7,4) BPSK with an general block interleaver [7:−1:1].

   a. Simulate the BER every dB between 0 and 22 dB with maximum Doppler shifts = 0.01 Hz and 0.1 Hz and compare the BER with theoretical uncoded BPSK

   b. Identify the model parameters in a. above

   c. Simulate the BER every dB between 0 and 12 dB with Hamming(7,4) coding and BPSK assuming a maximum Doppler shifts = 0.1 Hz and Alamouti 2 × 2 STBC; compare with theoretical uncoded BPSK for diversity = 4.

   d. Identify the model parameters in c. above

11

DIGITAL COMMUNICATIONS BER PERFORMANCE IN AWGN AND FADING (CONVOLUTIONAL CODING)

11.1 DIGITAL COMMUNICATIONS WITH CONVOLUTIONAL CODING IN AWGN AND FADING

This chapter presents topics in Simulink incorporating convolutional error control coding in an AWGN and fading channel. Specific topics include:

- BER performance of convolutional coding and BPSK in AWGN
  - Hard-decision decoding
  - Soft-decision decoding
- BER performance of convolutional coding and BPSK in AWGN and Rayleigh fading with interleaving
  - Hard-decision decoding
  - Soft-decision decoding
- BER performance of convolutional coding, BPSK and Alamouti STBC in Rayleigh fading with interleaving

---

Modeling of Digital Communication Systems Using SIMULINK®, First Edition.
Arthur A. Giordano and Allen H. Levesque.
© 2015 John Wiley & Sons, Inc. Published 2015 by John Wiley & Sons, Inc.
Companion Website: www.wiley.com/go/simulink

11.2 BER PERFORMANCE OF CONVOLUTIONAL CODING AND BPSK IN AWGN

11.2.1 Hard-Decision Decoding

In this section, Simulink models for convolutional codes with BPSK modulation in AWGN are presented. The BPSK demodulator is followed by a maximum likelihood decoder using the Viterbi algorithm (VA) where hard-decision decoding is selected. Simulink implements the convolutional encoder and decoder by means of a trellis structure for the generator polynomial with a specified constraint length and feedback taps given in octal. As an example, the poly2trellis(7, [171 133]) notation represents a trellis structure for a binary convolutional code with constraint length 7 and feedback taps located at the octal numbers 171 (binary 1111001) and 133(binary 1011011). Figure 11.1 shows the rate $1/2$ convolutional encoder for the poly2trellis(7, [171 133]) structure with one input and two outputs. The encoder has 6 delay elements denoted by Z^{-1} with the constraint length 7 corresponding to number of bits stored in the shift register. The free distance d_{free} of the convolutional code is the minimum distance in the set of all arbitrarily long paths that diverge from the all zero state and reenter the all zeros state.[1] The number of errors, t, that can be corrected by the code is given by[2]

$$t = \left\lfloor \frac{d_{\text{free}} - 1}{2} \right\rfloor$$

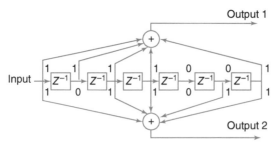

Figure 11.1 Rate $1/2$ Convolutional Encoder for Trellis Structure poly2trellis(7, [171 133]).

---

[1]The free distance is also described as the minimum Hamming distance between different encoded sequences of the same length. The Hamming distance corresponds to the number of positions in which the symbols differ.

[2]$\lfloor x \rfloor$ denotes the largest integer contained in x.

BER PERFORMANCE OF CONVOLUTIONAL CODING AND BPSK IN AWGN

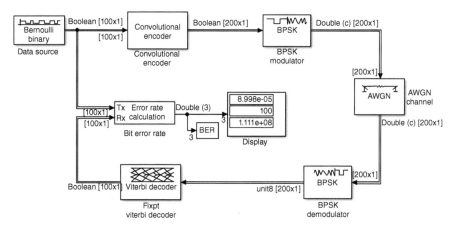

Figure 11.2 Simulink Model for Rate $1/2$ Convolutional Code and BPSK in AWGN with Hard Decisions.

The free distance for the convolutional code poly2trellis(7, [171 133]) is 10.

The Simulink model, depicted in Figure 11.2, is used to investigate two convolutional codes identified as follows:

1. Rate $1/2$ constraint length, K = 3, feedback taps [5 7] (poly2trellis(3, [5 7])), $d_{free} = 5$
2. Rate $1/2$ constraint length = 7, feedback taps [171 133] (poly2trellis(7, [171 133])), $d_{free} = 10$

In both cases, the Viterbi decoder is implemented with a traceback depth = 48 and hard decisions. The simulation is stopped once 100 errors are obtained.

Figure 11.2 shows that the Display block corresponds to (poly2trellis(3, [5 7])) with $E_b/N_o = 7$ dB.

The Simulink model parameters for the convolutional code (poly2trellis(3, [5 7])) are specified as follows:

Model Parameters for rate $1/2$, K = 3 Convolutional Code BPSK in AWGN

- BPSK antipodal signals = +1 and −1 (M = 2)
- Feedback taps [5 7]
- Symbol period = 0.5 s
- Frame-based with 100 samples/frame

- Sample time = 1 s
- Simulation time = stop with 100 errors
- Bernoulli binary probability of zero = 0.5
- Input signal power = 1 W
- Computation delay = Receive delay = 48 s
- Traceback depth = 48
- $E_s/N_o = E_b/N_o - 10\log(2)$
- AWGN with $\gamma_b = 7$ dB, hard decisions =>
    - simulated BER for rate $1/2$ convolutional code BPSK = 9×10^{-5}

The Simulink model parameters for the convolutional code poly2trellis(7, [171 133]) are specified as follows:

Model Parameters for Rate $1/2$, $K = 7$ Convolutional Code BPSK in AWGN

- BPSK antipodal signals = +1 and −1 ($M = 2$)
- Feedback taps [171 133]
- Symbol period = 0.5 s
- Frame-based with 100 samples/frame
- Sample time = 1 s
- Simulation time = stop with 100 errors
- Bernoulli binary probability of zero = 0.5
- Input signal power = 1 W
- Computation delay = Receive delay = 48 s
- Traceback depth = 48
- $E_s/N_o = E_b/N_o - 10\log(2)$
- AWGN with $\gamma_b = 6$ dB, hard decisions =>
    - simulated BER for rate $1/2$ convolutional code BPSK = 4.4×10^{-5}

11.2.1.1 Probability of Error for Hard-Decision Decoding of Convolutional Codes Using VA in AWGN The theoretical performance for hard-decision decoding of a convolutional code with a Viterbi decoder is available for comparison with simulation results. Assuming hard-decision decoding over a binary symmetric channel (BSC) with transition probability p, the probability of bit error is upper bounded by

$$P_e < \sum_{d=d_{\text{free}}}^{\infty} \beta_d P_2(d)$$

where $P_2(d)$ is given by

$$P_2(d) = \sum_{k=\frac{d}{2}+1}^{d} \binom{d}{k} p^k (1-p)^{d-k} + \frac{1}{2} \binom{d}{d/2} p^{d/2} (1-p)^{d/2}$$

and the coefficients β_d are related to the derivative of the transfer function of the code.[3] Note that for antipodal BPSK with coherent detection in AWGN

$$p = \frac{1}{2} \text{erfc} \left(\sqrt{\frac{E_b}{N_o}} \right)$$

Simulated BER performance for the two codes with coherent BPSK in AWGN is provided in Figure 11.3. Theoretical upper bounds obtained with the bertool are also depicted.

11.2.2 Soft-Decision Decoding

This section presents Simulink models for convolutional codes transmitted with BPSK modulation in AWGN assuming soft-decision Viterbi decoding. The Simulink model, depicted in Figure 11.4, is used to investigate two convolutional codes identified as follows:

1. Rate $1/2$, constraint length $K = 3$, feedback taps [5 7] (poly2trellis(3, [5 7])), $d_{\text{free}} = 5$
2. Rate $1/2$, constraint length $K = 7$, feedback taps [171 133] (poly2trellis(7, [171 133])), $d_{\text{free}} = 10$

In both cases, the Viterbi decoder is implemented with a traceback depth = 48 and 3-bit soft decisions. The simulation is stopped once 100 errors are obtained.

The principal difference between Simulink models shown in Figures 11.2 and 11.4 is the insertion of a quantizer block between the BPSK demodulator and the Viterbi decoder. An expanded view of the quantizer block, displayed by looking under the mask, is shown in Figure 11.4b where the output is

---

[3]Proakis, J.G., Digital Communications, 4th ed. McGraw-Hill 2001, pp. 489–490.

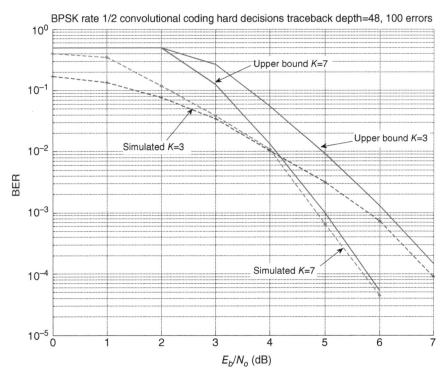

Figure 11.3 Simulated and Theoretical Upper Bounds of BER Performance for Rate $1/2$ Convolutional Codes with $K = 3$ and $K = 7$ Using Coherent BPSK.

represented as unit8 corresponding to unsigned 8-bit integers. Three-bit soft decisions are selected in the Viterbi soft-decision decoder.

The Simulink model parameters for the convolutional code poly2trellis(7, [171 133]) with soft decisions are specified as follows:

Model Parameters for rate $1/2$, $K = 7$ Convolutional Code BPSK in AWGN

- BPSK antipodal signals $= +1$ and -1 ($M = 2$)
- Feedback taps [171 133]
- $[-3:3]*1.9$ quantizer boundary points with output unit8
- Symbol period $= 0.5$ s
- Frame based with 100 samples/frame
- Sample time $= 1$ s
- Simulation time $=$ stop with 100 errors

BER PERFORMANCE OF CONVOLUTIONAL CODING AND BPSK IN AWGN

- Bernoulli binary probability of zero = 0.5
- Input signal power = 1 W
- BPSK soft demod noise variance = $1/(10^{\wedge}((E_bN_o-10*\log10(2))/10))$
- Computation delay = receive delay = 48 s
- Traceback depth = 48
- $E_s/N_o = E_b/N_o - 10\log(2)$
- AWGN with $\gamma_b = 4$ dB, 3-bit soft decisions =>
  - simulated BER for rate 1/2 convolutional code BPSK = 3×10^{-5}

The Simulink model parameters for the convolutional code (poly2trellis(3, [5 7])) with soft decisions are specified as follows:

Model Parameters for rate 1/2, $K = 3$ Convolutional Code BPSK in AWGN

- BPSK antipodal signals = +1 and −1 ($M = 2$)
- Feedback taps [5 7]
- [−3:3]*1.9 quantizer boundary points with output unit8
- Symbol period = 0.5 s
- Frame based with 100 samples/frame
- Sample time = 1 s
- Simulation time = stop with 100 errors
- Bernoulli binary probability of zero = 0.5
- Input signal power = 1 W
- BPSK soft demod noise variance = $1/(10^{\wedge}((E_bN_o-10*\log(2))/10))$
- Computation delay = receive delay = 48 s
- Traceback depth = 48
- $E_s/N_o = E_b/N_o - 10\log(2)$
- AWGN with $\gamma_b = 4$ dB, 3-bit soft decisions =>
  - simulated BER for rate 1/2 convolutional code BPSK = 8.8×10^{-4}

11.2.2.1 Probability of Error for Soft-Decision Decoding of Convolutional Codes Using VA in AWGN[5] The theoretical performance for soft-decision decoding of a convolutional code with a Viterbi decoder is

---

[4]The Soft decision model was provided by The MathWorks™.
[5]Proakis, J.G., Digital Communications, 4th ed. McGraw-Hill 2001, pp. 485–491.

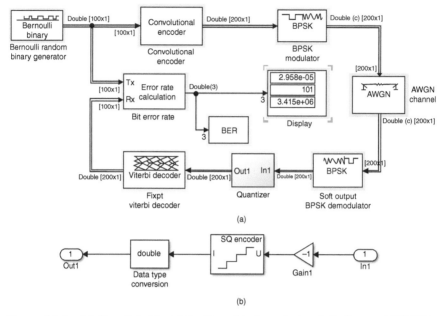

Figure 11.4 (a) Simulink Model for Rate $1/2$ Convolutional Coding and BPSK in AWGN with Soft-Decision Decoding; (b) Expanded View of the Quantizer Block.[4]

available for comparison. Assuming that the all-zeros code word is sent, a variable a_d is defined to be the number of paths of distance d from the all zero path that merges with the all zero path for the first time. It is related to the variable $\beta_d = a_d f(d)$, where $f(d)$ is a function of the distance. An upper bound on the bit error probability is then expressed as

$$P_b < \sum_{d=d_{\text{free}}}^{\infty} \frac{\beta_d}{2} \text{erfc}\left(\sqrt{R_c d \gamma_b}\right) < \sum_{d=d_{\text{free}}}^{\infty} \beta_d e^{-R_c d \gamma_b}$$

Simulated BER performance in AWGN for the two codes with coherent BPSK is provided in Figure 11.5. Theoretical upper bounds obtained with the bertool are also depicted.

Figure 11.6 compares BER results for hard and soft decisions using the rate $1/2$ $K = 7$ convolutional code. It is observed that there is about a 2 dB gain for soft over hard decisions.

In AWGN, the coding gain in dB for soft-decision decoding compared to uncoded coherent BPSK is bounded by $10 \log(R d_{\text{free}})$. For example, for $R = 1/2$, $K = 7$ with $d_{\text{free}} = 10$, the coding gain is bounded by $10\log(10/2) = 7$ dB. From Figure 11.7, it is observed that for a BER $\sim 2 \times 10^{-5}$ the coding gain is a little over 5 dB.

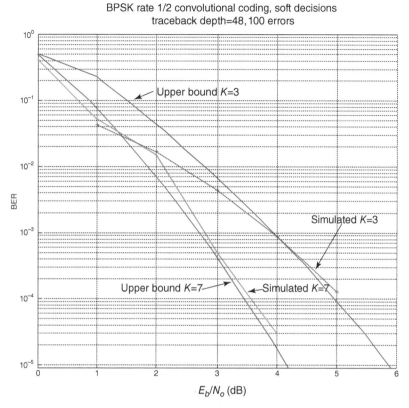

Figure 11.5 BER Performance of Rate $1/2$ Convolutional Code and BPSK in AWGN with Soft Decisions and Constraint Lengths $K = 3$ and 7.

11.3 BER PERFORMANCE OF CONVOLUTIONAL CODING AND BPSK IN AWGN AND RAYLEIGH FADING WITH INTERLEAVING (SOFT- AND HARD-DECISION DECODING)

In this section, Simulink models and BER performance are obtained for rate $1/2$ $K = 7$ convolutional coding and BPSK in Rayleigh fading. Due to the fading behavior, the models must include an interleaver to disperse error bursts allowing the code to be effective in correcting the errors. The Simulink models presented for hard decisions in Figure 11.2 and soft decisions in Figure 11.4 are expanded to include matrix interleaving and deinterleaving as shown in Figure 11.8 for hard decisions and in Figure 11.9 for soft decisions. The input to the matrix interleaver is entered row-by-row and its output is produced column-by-column; the number of rows and columns are each 14 for the models as shown in Figures 11.8 and 11.9. The deinterleaver performs the inverse

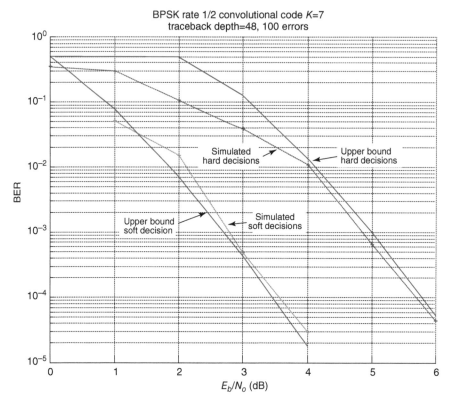

Figure 11.6 Comparison of BER Performance of Rate $1/2$ Convolutional Code and BPSK in AWGN with Hard and Soft Decisions and Constraint Length $K = 7$.

operation reading data in column-by-column and outputting data row-by-row. The scopes shown in Figure 11.8 are used to verify that the transmitted and received signals are synchronized due to a delay introduced by the presence of the interleaver/deinterleaver combination. For comparison of simulation results both AWGN and Rayleigh fading channel models are included.

The Simulink model parameters for the model shown in Figure 11.8 are specified as follows:

Model Parameters for rate $1/2$, $K = 7$ Convolutional Code BPSK in Rayleigh Fading and AWGN (Hard Decisions)

- BPSK antipodal signals $= +1$ and -1 ($M = 2$)
- Feedback taps [171 133]
- Symbol period $= 0.5$ s
- Frame based with 98 samples/frame

BER PERFORMANCE OF CONVOLUTIONAL CODING

- Simulation time = stop with 200 errors
- Bernoulli binary probability of zero = 0.5
- Input signal power = 1 W
- Delay for error rate computation = computation delay = 48 s
- Jakes fading with maximum Doppler shift = 0.1 Hz
- Interleaver 14×14
- Receive delay = 0 s
- Traceback depth = 48
- $E_s/N_o = E_b/N_o - 10\log(2)$
- Rayleigh fading with $\gamma_b = 7$ dB, hard decisions =>
  - simulated BER for rate $1/2$ convolutional code BPSK = 0.054
- AWGN with $\gamma_b = 4$ dB, hard decisions =>
  - simulated BER for rate $1/2$ convolutional code BPSK = 8.5×10^{-3}

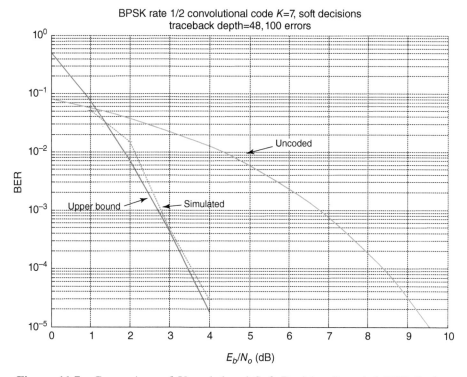

Figure 11.7 Comparison of Uncoded and Soft Decision Decoded BER Performance of Rate $1/2$ Convolutional Code and BPSK in AWGN with Constraint Length $K = 7$.

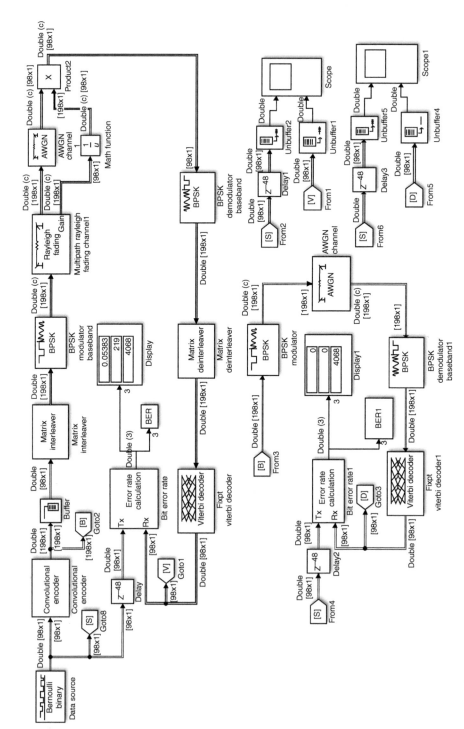

Figure 11.8 Simulink Model for Rate ½ Convolutional Code and BPSK in Rayleigh Fading with Hard Decisions.

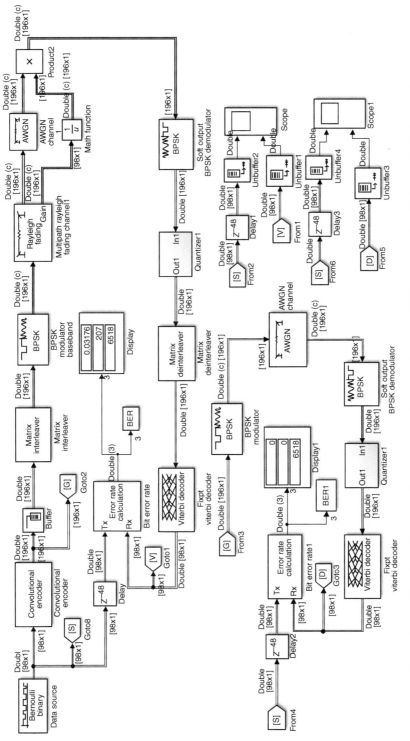

Figure 11.9 Simulink Model for Rate ½ Convolutional Code and BPSK in Rayleigh Fading with Soft Decisions.

The Simulink model parameters for Figure 11.9 are specified as follows:

Model Parameters for rate $1/2$, $K = 7$ **Convolutional Code BPSK in Rayleigh Fading and AWGN (Soft Decisions)**

- BPSK antipodal signals $= +1$ and -1 ($M = 2$)
- Feedback taps [171 133]
- Symbol period $= 0.5$ s
- Frame based with 98 samples/frame
- Simulation time $=$ stop with 200 errors
- Bernoulli binary probability of zero $= 0.5$
- Input signal power $= 1$ W
- BPSK soft demod noise variance $= 1/(10^{\wedge}((E_b N_o - 10*\log(2))/10))$
- Delay for error rate computation $=$ computation delay $= 48$ s
- Jakes fading with maximum Doppler shift $= 0.1$ Hz
- Interleaver 14×14
- Receive delay $= 0$ s
- Traceback depth $= 48$
- $E_s/N_o = E_b/N_o - 10\log(2)$
- $[-3:3]*1.9$ quantizer boundary points with output unit8
- Rayleigh fading with $\gamma_b = 7$ dB, 3-bit soft decisions $=>$
  - simulated BER for rate $1/2$ convolutional code BPSK $= 0.032$
- AWGN with $\gamma_b = 5$ dB, hard decisions $=>$
  - simulated BER for rate $1/2$ convolutional code BPSK $= 2.8 \times 10^{-5}$

Figure 11.10 presents the BER performance for rate $1/2$ $K = 7$ convolutional coding with BPSK for hard- and soft-decision decoding. Note that only the Jakes model and a single value of 0.1 Hz maximum Doppler are selected for demonstration purposes. The simulated results illustrate the degradation in performance due to Rayleigh fading versus AWGN transmission.

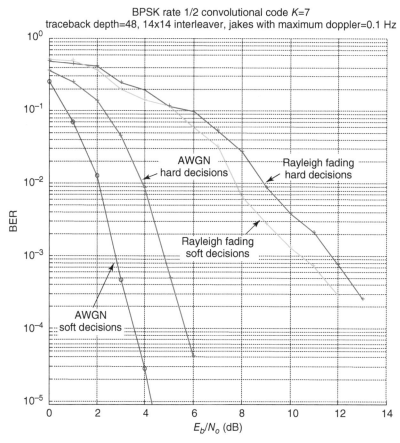

Figure 11.10 BER Performance of Rate $1/2$ Convolutional Code and BPSK with $K = 7$ in Rayleigh Fading with Hard- and Soft-Decisions.

11.4 BER PERFORMANCE OF CONVOLUTIONAL CODING AND BPSK AND ALAMOUTI STBC IN RAYLEIGH FADING WITH INTERLEAVING

This remaining section presents Simulink models and BER performance for rate-$1/2$ $K = 7$ convolutional coding and BPSK in Rayleigh fading with Alamouti STBC. Figure 11.11 expands Figure 11.8 for hard decisions and Figure 11.12 expands Figure 11.9 for soft decisions to incorporate Alamouti 2×2 STBC. Figure 11.13 depicts the 2×2 Rayleigh fading channel model used for both hard- and soft-decision Simulink models.

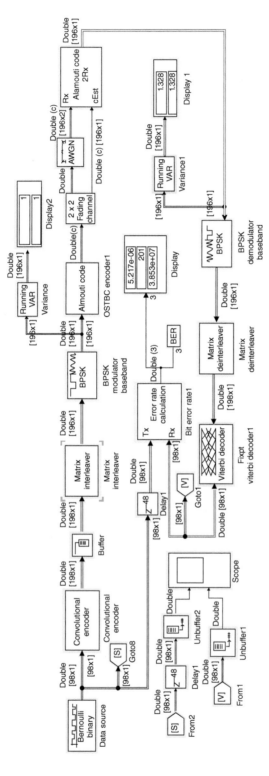

Figure 11.11 Simulink Model for Rate 1/2 Convolutional Code and BPSK in Rayleigh Fading with Hard Decisions and Alamouti STBC 2×2.

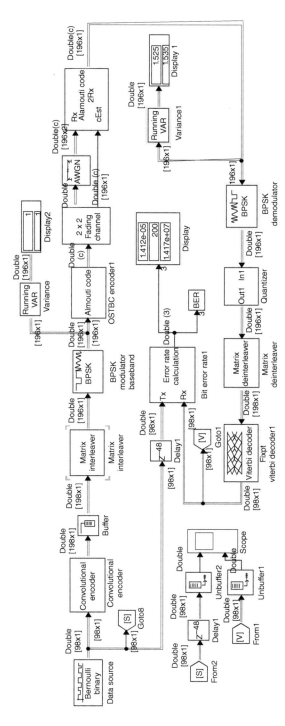

Figure 11.12 Simulink Model for Rate $1/2$ Convolutional Code and BPSK in Rayleigh Fading with Soft Decisions and Alamouti STBC 2×2.

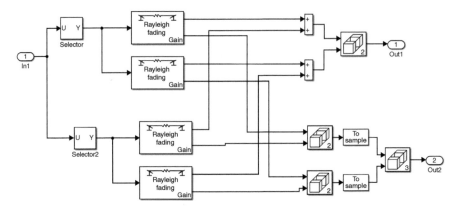

Figure 11.13 Simulink Model for Rayleigh Fading and Alamouti STBC 2×2.

The Simulink model parameters for Figures 11.11 and 11.12 are specified as follows:

Model Parameters for rate 1/2, K = 7 Convolutional Code BPSK in Rayleigh Fading with AWGN and Alamouti STBC 2×2

- BPSK antipodal signals = +1 and −1 ($M = 2$)
- Feedback taps [171 133]
- Symbol period = 0.5 s
- Frame based with 98 samples/frame
- Simulation time = stop with 200 errors
- Bernoulli binary probability of zero = 0.5
- Input signal power = 2 W
- Delay for error rate computation = computation delay = 48 s
- Jakes fading with maximum Doppler shift = 0.1 Hz
- Interleaver 14 × 14
- Receive delay = 0 s
- Traceback depth = 48
- 3-bit soft decisions
- [−3:3]*1.9 quantizer boundary points with output unit8
- $E_b N_o - 10*\log(2) + 10*\log(1/2)$
- BPSK soft demod noise variance
  = $1/(10^{\wedge}((E_b N_o - 10*\log(2)+10*\log(1/2))/10))$

SUMMARY DISCUSSION 243

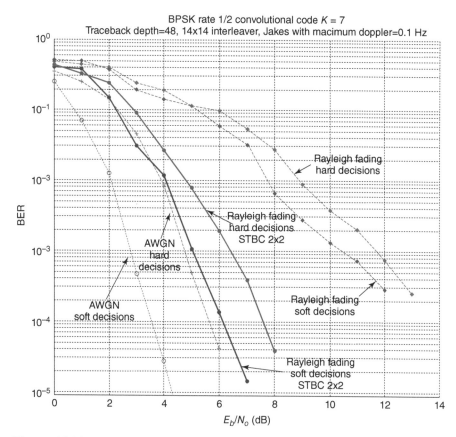

Figure 11.14 BER Performance of Rate $1/2$ Convolutional Code and BPSK for $K = 7$ in Rayleigh Fading with Hard and Soft Decisions and Alamouti STBC 2×2.

Figure 11.14 presents the BER performance in Rayleigh fading for rate $1/2$ $K = 7$ convolutional coding with BPSK for hard- and soft-decision decoding using Alamouti 2×2 STBC. The BER results indicate that 2×2 Alamouti STBC recovers much of the performance loss caused by Rayleigh fading.

11.5 SUMMARY DISCUSSION

This chapter has presented BER results using the VA in the receiver for various combinations of convolutional codes and modulations over AWGN and fading channels. Simulated BER results were obtained for hard- and soft-decision decoding and found to compare favorably with upper bounds in

AWGN channels. In Rayleigh fading channels, the use of interleaving allows the convolutional error control codes to utilize their free distance properties to correct bit errors and provide significant coding gain over uncoded schemes that use the same modulation. The last section of this chapter repeated the convolutional error control coding and modulation cases with interleaving and Alamouti STBC. Incorporation of STBC indicates that much of the performance loss due to Rayleigh fading is recovered.

PROBLEMS

11.1 In the Simulink model shown in Figure 11.4 for the convolutional code with $K = 7$ and soft decisions, change the traceback depth from 48 to 32 and 10 and determine BER for E_b/N_o in the range from 0 to 4 dB with 100 errors for all three cases of traceback depth.

11.2 Find the simulated BER for the convolutional code $K = 9$ with feedback taps [561 753] for E_b/N_o in the range from 0 to 5 dB. Assume AWGN with coherent BPSK, the traceback depth = 56 and hard decisions. List the Simulink model parameters and determine the upper bound on post-decoding BER from the bertool.

11.3 Using the Simulink model shown in Figure 11.9 for soft decisions and the associated Simulink model parameters, obtain the BER in Rayleigh fading with a stop criterion of both 100 and 200 errors for E_b/N_o values between 0 and 12 dB in 1 dB steps. List the Simulink model parameters.

11.4 Using the Simulink model shown in Figure 11.9 for soft decisions and the associated Simulink model parameters, obtain the BER in Rayleigh fading with a stop criterion of 200 errors for E_b/N_o values between 0 and 12 dB in 1 dB steps with both 0.1 and 0.01 maximum Doppler shift. List the Simulink model parameters.

11.5 Modify the Simulink models shown in Figures 11.8 and 11.9 to incorporate 28×28 interleaving. Obtain the BER for hard- and soft-decisions in Rayleigh fading and compare the results with 14×14 interleaving. Display the Simulink models for hard- and soft-decisions. List the Simulink model parameters.

11.6 Using the Simulink model in Figure 11.9 for soft decisions and the associated Simulink model parameters, obtain the BER with a stop criterion of 200 errors for E_b/N_o values between 0 and 10 dB in 1 dB steps where the Rayleigh fading channel is replaced with a Rician channel using the

PROBLEMS

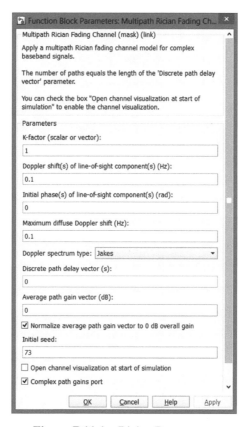

Figure P.11.1 Rician Parameters.

Rician parameters shown in Figure P.11.1. Show the model and plot the BER for both the Rayleigh and Rice cases.

11.7 Replace the Simulink models in Figures 11.11 and 11.12 with Alamouti STBC 2×1. Compute the BER for this case and compare the results with STBC 2×2. List the Simulink model parameters and display the hard- and soft-decision models including the STBC 2×1 model.

12

ADAPTIVE EQUALIZATION IN DIGITAL COMMUNICATIONS

12.1 ADAPTIVE EQUALIZATION

This chapter presents Simulink models that incorporate adaptive equalization. Specific topics include the following:

- Linear least mean square (LMS) equalizers with BPSK in dispersive multipath channels and AWGN including
  - Sample based
  - Frame based
  - Simulink library model
- Linear LMS equalizer with QPSK in a channel with Intersymbol Interference (ISI)
- Decision feedback equalizers (DFEs) with BPSK in a dispersive multipath channel and AWGN
- Recursive least squares (RLS) equalizers with BPSK in a dispersive multipath channel, AWGN, and Rayleigh fading

---

Modeling of Digital Communication Systems Using SIMULINK®, First Edition.
Arthur A. Giordano and Allen H. Levesque.
© 2015 John Wiley & Sons, Inc. Published 2015 by John Wiley & Sons, Inc.
Companion Website: www.wiley.com/go/simulink

12.2 BER PERFORMANCE OF BPSK IN DISPERSIVE MULTIPATH CHANNEL USING AN LMS LINEAR EQUALIZER

Equalization algorithms are used for mitigating ISI and/or multipath introduced by the channel. The channel is modeled as a discrete-time channel filter with delay values corresponding to the symbol interval T and channel coefficients that represent the channel distortion. The channel model seen in Figure 12.1 includes AWGN. In the figure, the data signal symbols from the BPSK modulator are identified by the sequence $\{s_k\}$ and are corrupted by the channel coefficients $\{a_m\}$. The received signal $\{r_k\}$ is then expressed in terms of the channel coefficients as

$$r_k = \sum_{m=0}^{L-1} a_m s_{k-mT} + n_k \quad k = 0, 1, \ldots$$

where $\{n_k\}$ represents zero mean AWGN with variance σ^2 and L is the number of tap coefficients in the channel.

Dispersive multipath channels or channels that cause ISI generally result in unacceptable BER. Adaptive equalizers compensate for these channel distortions and provide significant improvement in BER. This section presents linear LMS equalizers implemented using a finite duration impulse response (FIR) filter with delay elements corresponding to the symbol duration and coefficients that are adaptively determined by an LMS algorithm.[1] Figure 12.2 depicts a linear transversal equalizer with $2M+1$ equalizer coefficients, $\{c_i, i = -M \ldots M\}$, where the number of taps in the equalizer is selected to span the multipath spread or a finite number of ISI symbols. The output of the equalizer is the estimate of the kth symbol $\hat{s}_k$.

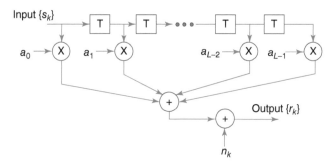

Figure 12.1 Discrete Time Channel Model.

---

[1] More generally, the equalizer delay elements may be fractionally spaced.

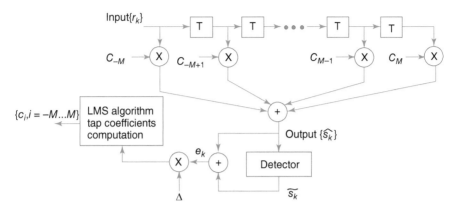

Figure 12.2 LMS Equalizer.

An estimate of the transmitted signal is now expressed as

$$\widehat{s}_k = \sum_{i=-M}^{M} c_i r_{k-iT} \quad k = 0, 1, \ldots$$

An error e_k, occurs if s_k is unequal to $\widehat{s}_k$. Assuming p is an integer that controls the tap for synchronizing the received signal, the error signal is given by $e_k = s_{k-pT} - \widehat{s}_k$. In this notation, the desired signal, $d_k = s_{k-pT}$, is a delayed version of the BPSK modulated symbols and is used for training the equalizer with known symbols at the receiver. In some equalizer designs, known symbols are multiplexed with unknown symbols.

The coefficients of the LMS algorithm are determined by minimizing the mean square error (MSE) defined as an ensemble average $E\{|e_i|^2\}$. The minimization is accomplished by applying the orthogonality principle requiring that the error be orthogonal to the input signal sequence,[2] that is,

$$E\{e_k r^*_{k-lT}\} = 0, \quad -M \le l \le M$$

Incorporating the received signal in the aforementioned relation results in

$$E\{r^*_{k-lT} s_{k-pT}\} = \sum_{m=-M}^{M} c_m E\{r^*_{k-lT} r_{k-mT}\}$$

---

[2]Schonhoff, T.A., and A.A. Giordano, op.cit., pp. 385–391.

The term on the left side of the aforementioned equation is the cross correlation between the transmitted and received signal where the term in brackets on the right side of the aforementioned equation is the autocorrelation of the received signal. In matrix form this equation is known as the normal equations and is expressed as

$$AC = G$$

where A is the autocorrelation matrix of the received signal, C is the vector of equalizer coefficients, and G is the cross correlation matrix formed between the transmitted and received signals.

For clarity, it is helpful to present a simple analytical example prior to proceeding with the Simulink model. Consider an example where the channel is a two-path multipath channel with real coefficients $a_0 = 1$ and $a_1 = a$.[3] The data symbols are assumed to be uncorrelated, with equal probability and have values $+$ and -1. Then $E\{s_k\} = 0$ and $E\{s_k^2\} = 1$. In addition, assume that the equalizer has three coefficients c_{-1}, c_0, and c_1. Then the orthogonality principle is stated as

$$E\left\{r_{k-lT}^*\left[s_{k-pT} - \sum_{m=-1}^{1} c_m r_{k-mT}\right]\right\} = 0 \quad \text{for} \quad -1 \leq l \leq 1$$

Using $r_{k-lT}^* = s_{k-lT}^* + a s_{k-(l+1)T}^* + n_{k-lT}^*$, the moments are now computed as follows:

$$E\left\{r_{k-lT}^* s_{k-pT}\right\} = E\left\{s_{k-pT}\left[s_{k-lT}^* + a s_{k-(l+1)T}^* + n_{k-lT}^*\right]\right\} = \begin{cases} a & l = p-1 \\ 1 & l = p \\ 0 & \text{otherwise} \end{cases}$$

$$E\left\{r_{k-lT}^* r_{k-mT}\right\} = \begin{cases} 1 + a^2 + \sigma^2 & l = m \\ a & l = m-1, \; l = m+1 \\ 0 & \text{otherwise} \end{cases}$$

---

[3] Schonhoff, T.A., and A.A. Giordano, op.cit., pp. 397–400.

In matrix form with $p=0$, the aforementioned expressions become

$$\begin{bmatrix} (1+|a|^2)+\sigma^2 & a & 0 \\ a & (1+|a|^2)+\sigma^2 & a \\ 0 & a & (1+|a|^2)+\sigma^2 \end{bmatrix} \begin{bmatrix} c_{-1} \\ c_0 \\ c_1 \end{bmatrix} = \begin{bmatrix} a \\ 1 \\ 0 \end{bmatrix}$$

Ignoring the noise and letting $a = 0.5$, the coefficients can be computed, that is,

$$\begin{bmatrix} c_{-1} \\ c_0 \\ c_1 \end{bmatrix} = \begin{bmatrix} 0.02353 \\ 0.9412 \\ -0.3765 \end{bmatrix}$$

A Simulink model for this example is shown in Figure 12.3. The results indicate that no errors occur using the three coefficients computed above.

The fixed channel coefficients assumed earlier ordinarily apply for a limited interval. In general, the equalizer must track the time-varying channel characteristics. In this case, the equalizer coefficients must adapt to the time-variation. An algorithm that accomplishes this task for channels with slow time variations is referred to as a stochastic gradient algorithm and is more commonly referred to as an LMS algorithm. The equalizer coefficients are computed iteratively to determine the coefficients according to

$$\hat{c}_{k+1} = \hat{c}_k + \Delta e_k r_k^*$$

where Δ is a scale factor that controls the rate of convergence. The error signal may be determined by forming the difference between the detector output, $\tilde{s}_k$, and the equalizer output, $\hat{s}_k$, as shown in Figure 12.2, or by use of a training sequence that is known at the receiver and is multiplexed with the transmitted signal. A Simulink model that implements a three tap LMS adaptive equalizer using a training signal is shown in Figure 12.4 where the two-path multipath channel is again utilized. The received signal is synchronized to the center tap of the equalizer.

Figure 12.5 is an expanded view of the tap coefficient implementation and includes a thirty-first order FIR filter. The magnitude of the filter response is shown in Figure 12.6 along with parameters chosen for the simulation.[4]

---

[4] See Mathworks Filter Design and Analysis Tool (fdatool).

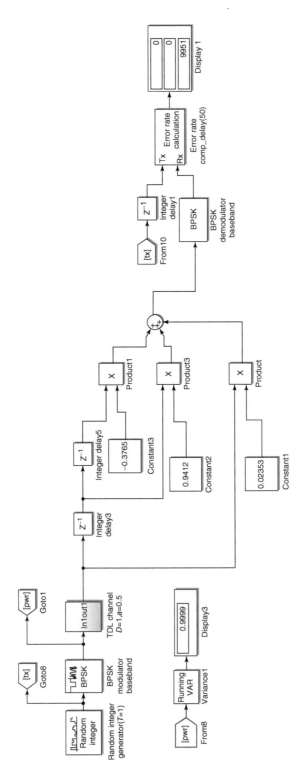

Figure 12.3 Simulink Model for LMS Equalizer with Three Coefficients.

252

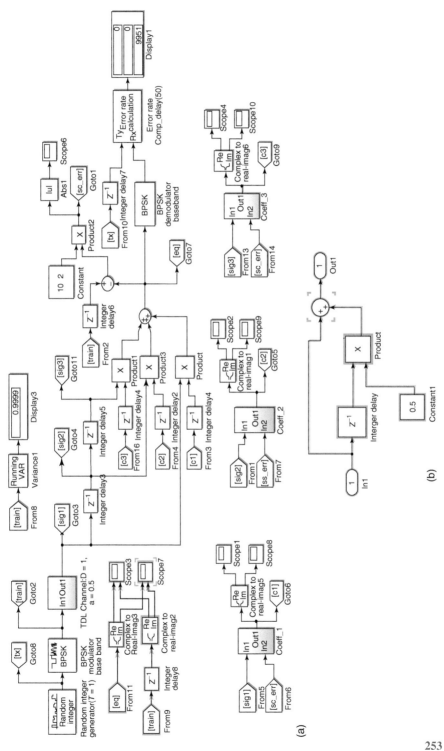

Figure 12.4 (a) Simulink Model for Adaptive LMS Equalizer with Three Coefficients; (b) Two-Path Multipath Model with $a = 0.5$, Delay = 1 s.

254 ADAPTIVE EQUALIZATION IN DIGITAL COMMUNICATIONS

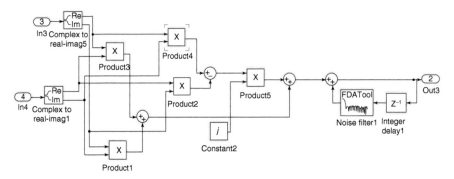

Figure 12.5 Coefficient Computation using Thirty-First Order FIR Filter.

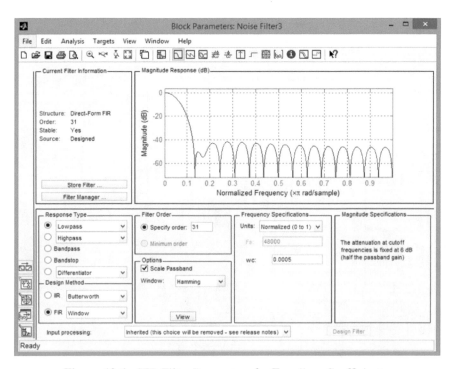

Figure 12.6 FIR Filter Parameters for Equalizer Coefficients.

The Simulink model parameters for this example are specified as follows:

Model Parameters for 3 Coefficient LMS Adaptive Equalizer

- BPSK antipodal signals $= +1$ and -1 ($M=2$)
- Symbol period $= 1$ s
- Sample based with 1 sec sample time
- Simulation time $= 10{,}000$ s
- Input signal power $= 1$ W
- Channel: Two-path multipath $(1, 0.5)$ with 1 s delay and no noise
- Scale factor $\Delta = 0.01$
- Synchronize received signal to center tap
- Ideal equalizer training
- No errors by end of simulation

Execution of the Simulink model shown in Figure 12.4 produces the following results:

- Figure 12.7 depicts a portion of the equalizer output signal along with the transmitted signal. This figure indicates that in this noiseless case a detector would make the correct decisions, that is, no errors with equalization, even though the equalized signal is a distorted version of the transmitted signal.
- The magnitude of the error signal versus time, observed in Figure 12.4a (scope6), is depicted in Figure 12.8.
- Figure 12.9 depicts the convergence of each of the three LMS equalizer coefficients. The real parts of the coefficients are obtained from scope1, scope2, and scope3 shown in Figure 12.3; (The corresponding imaginary parts are all zero.)

The converged values of the three coefficients $\{c_1, c_2, c_3\}$ in Figure 12.9 are seen to be very close to the theoretical values $\{c_{-1}, c_0, c_1\}$ computed earlier.

In this simulation, the training sequence utilized was the data from the BPSK modulator. In an actual implementation, two common schemes include: (i) using an embedded training signal known at both the transmitter and receiver and (ii) subtracting the detected symbols from the equalizer output to compute the error sequence. Therefore, the results provided here

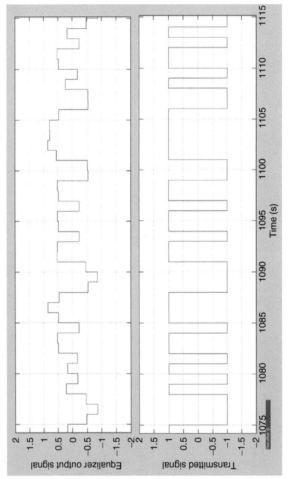

Figure 12.7 A Sample of the Equalizer Output (Top Trace) and Transmitted Signal (Bottom Trace).

LMS LINEAR EQUALIZER FROM THE SIMULINK LIBRARY

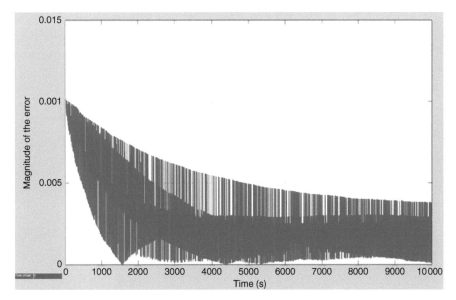

Figure 12.8 Magnitude of LMS Error.

are ideal but useful in illustrating the use of Simulink in modeling an adaptive system.

12.3 BER PERFORMANCE OF BPSK IN DISPERSIVE MULTIPATH CHANNEL USING AN LMS LINEAR EQUALIZER FROM THE SIMULINK LIBRARY

As shown in Figure 12.10 the LMS equalizer from the Simulink library is used to equalize the two-path multipath channel discussed in the previous section. The equalizer parameters, shown in Figure 12.11, indicate three equalizer coefficients, synchronization using the center tap and a 0.01 scale factor.

The Simulink model parameters for this example are specified as follows:

Model Parameters for 3 Coefficient LMS Adaptive Equalizer from the Simulink Library

- BPSK antipodal signals $= +1$ and -1 ($M = 2$)
- Symbol period $=$ sample time $= 1$ s

- Frame based with 10 samples/frame
- Simulation time = 10,000 s
- Input signal power = 1 W
- Channel: two-path multipath (1, 0.5) with 1 s delay
- Scale factor $\Delta = 0.01$
- Synchronize received signal to center tap
- Number of sample/symbol = 1
- Leakage factor = 1
- Initial weights = 0
- Ideal equalizer training
- $E_b/N_o = 10$ dB

The Simulink model shown in Figure 12.10 is executed with E_b/N_o in the AWGN block specified to be 10 dB. As a result, Figure 12.12 displays the following signals:

- the signal constellations at the BPSK modulator output (Figure 12.12a)
- the equalizer input (Figure 12.12b), and
- the equalizer output (Figure 12.12c)

The cluster of signal samples shown in Figure 12.12b is suppressed by the equalizer as seen in Figure 12.12c.

Figure 12.13 provides a bench mark for ideal BPSK BER assuming AWGN only. With the multipath channel, the degradation from ideal BPSK BER performance is apparent. Substantial improvement in BER is then attained by use of the adaptive equalizer.

12.4 BER PERFORMANCE OF QPSK IN A CHANNEL WITH ISI USING AN LMS LINEAR EQUALIZER

In this section, a QPSK modulated signal is corrupted by AWGN and a digital filter that introduces ISI as shown in the Simulink model in Figure 12.14.

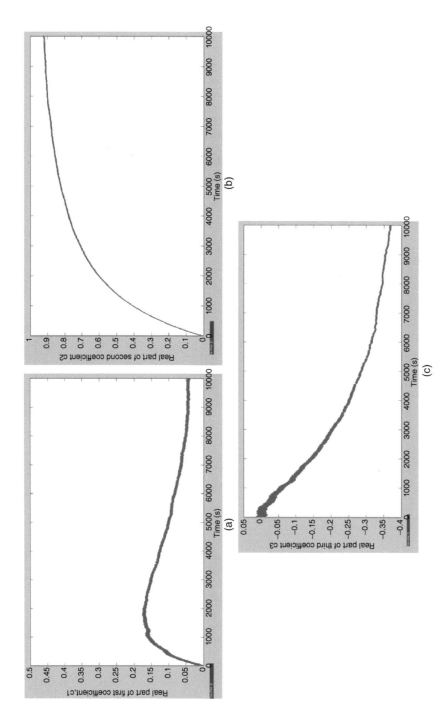

Figure 12.9 (a) Real Part of Coefficient LMS Equalizer c_1; (b) Real Part of Coefficient LMS Equalizer c_2; (c) Real Part of Coefficient LMS Equalizer c_3.

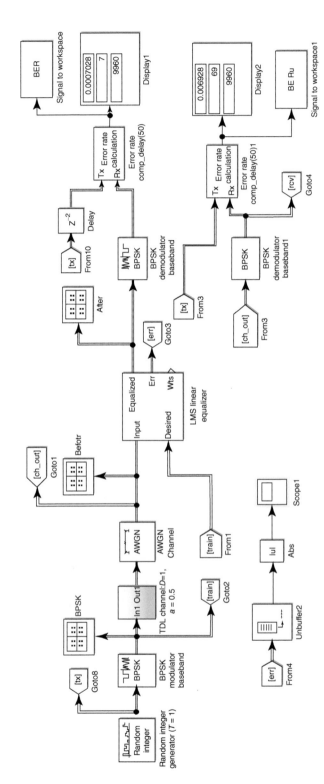

Figure 12.10 LMS Equalizer with Three Coefficients from the Simulink Library ($E_b/N_o = 10$ dB).

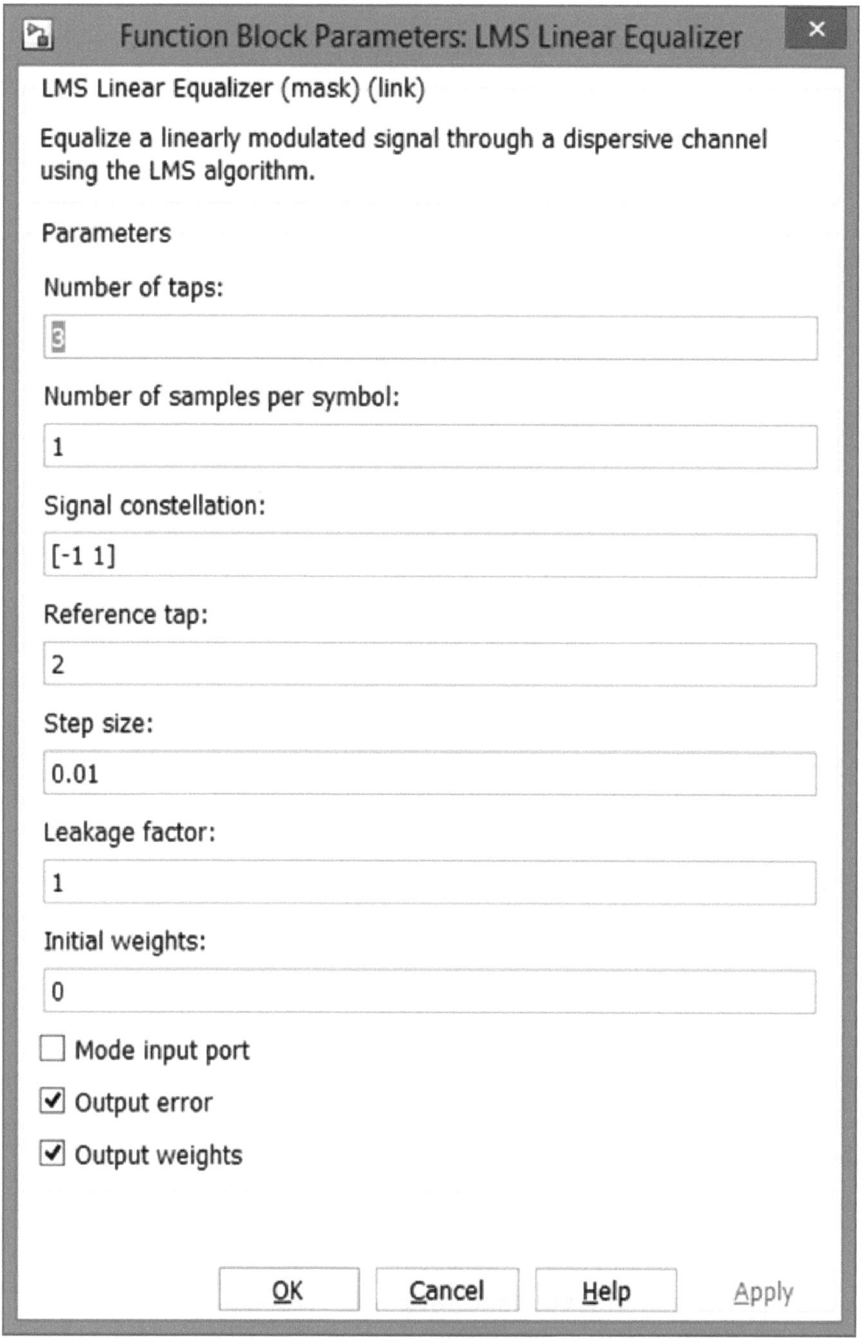

Figure 12.11 Parameter Choices for LMS Equalizer from Simulink Library.

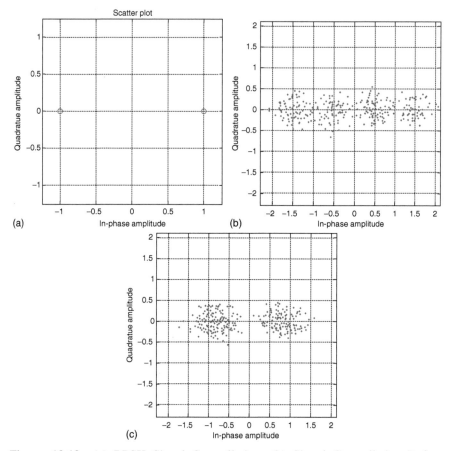

Figure 12.12 (a) BPSK Signal Constellation; (b) Signal Constellation Before the Equalizer with E_b/N_o = 10 dB; (c) Signal Constellation After Adaptive Equalization.

The Simulink model parameters for this example are specified as follows:

Model Parameters for LMS Adaptive Equalizer with Digital Filter

- QPSK signals ($M = 4$) Gray encoding
- Symbol period = sample time = 0.001 s
- Frame based with 1000 samples/frame

BER PERFORMANCE OF QPSK IN A CHANNEL WITH ISI

- Simulation time = 1000 s
- Input signal power = 1 W
- FIR digital filter with numerator coefficients [1 −0.3 0.1 0.2j]
- 6 tap LMS linear equalizer
- LMS equalizer signal constellation = cos(pi/4)*qammod([0:3],4)
- Scale factor $\Delta = 0.001$
- Synchronize received signal to tap 2 (reference tap)
- Number of samples/symbol = 1
- Leakage factor = 1
- Initial weights = 0
- Receive delay = computation delay = 0 s
- Ideal equalizer training
- $E_s/N_o = 6$ dB => probability of symbol error = 0.047 (simulated)

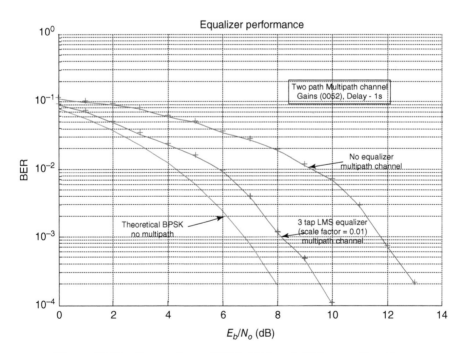

Figure 12.13 Comparison of BERs With and Without an Equalizer.

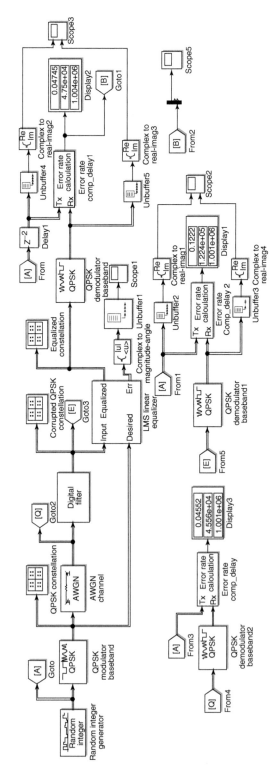

Figure 12.14 Simulink Model for QPSK with AWGN and ISI Using LMS Equalizer ($E_s/N_o = 6\,\text{dB}$).

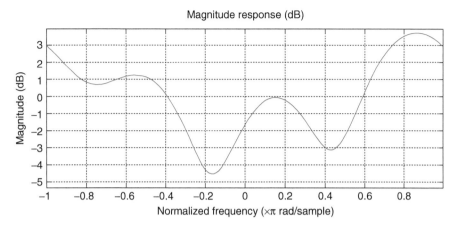

Figure 12.15 Digital Filter Frequency Response.

The digital filter, a copy of the filter in the Mathworks Simulink model `doc_lmseq`, is an all-zeros FIR filter with numerator coefficients [1 −0.3 0.1 0.2j] and a frequency response shown in Figure 12.15.[5]

Figure 12.16 depicts the QPSK signal constellation with signal values at

$$-0.7071 + 0.7071j, -0.7071 - 0.7071j, 0.7071 + 0.7071j,$$
$$0.7071 - 0.7071j$$

Figure 12.17 shows the QPSK scatter diagrams following the AWGN and digital filter with two values of E_s/N_o. Figure 12.18 depicts the QPSK scatter diagrams after equalization with the two E_s/N_o values where the clustering around the transmitted signal values is more evident with the higher value of E_s/N_o. With $E_s/N_o = 50$ dB Figure 12.19 shows rapid convergence of the equalized error magnitude.

Figure 12.20 illustrates that the digital filter does not introduce any delay and results in the same demodulated signal as the source with $E_s/N_o = 50$ dB. The top trace shown in Figure 12.20 is the real part of the source output; the bottom trace is the real part of the QPSK demodulator output with the digital filter and AWGN. In this case, the ISI does not affect the symbol error probability.

Figure 12.21 illustrates that the equalizer is synchronized with the transmitted signal by selecting the tap 2 reference and stipulating that the signal

---

[5]See MathWorks documentation in Communications System Toolbox/System Design/Equalization entitled Implement LMS Linear Equalizer Using Simulink. A Simulink example that uses the Digital Filter block with 16-QAM is `doc_lmseq`.

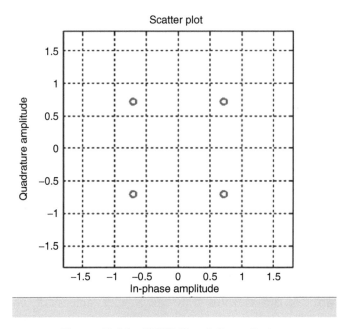

Figure 12.16 QPSK Signal Constellation.

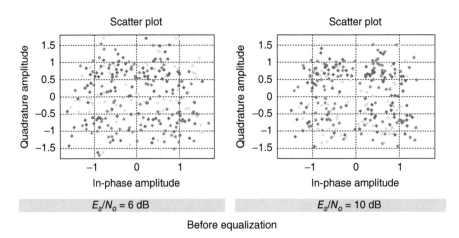

Before equalization

Figure 12.17 QPSK Scatter Diagrams Following the AWGN and Digital Filter.

from the source be delayed by two symbols for the error probability computation. In this figure, the top trace is the real part of the source output with a two symbol delay; the bottom trace displays the equalizer output for $E_s/N_o = 50$ dB where errors are apparent during the equalizer convergence interval.

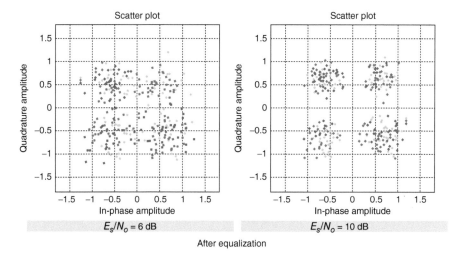

Figure 12.18 QPSK Scatter Diagrams After Equalization.

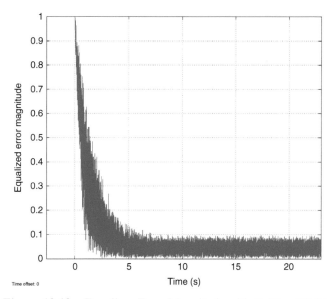

Figure 12.19 Equalizer Error Magnitude with $E_s/N_o = 50$ dB.

Figure 12.22 shows that the probability of symbol error with $E_s/N_o = 10$ dB decreases as the equalizer converges.

Table 12.1 presents symbol error probabilities for selected values of E_s/N_o. Note that with $E_s/N_o = 50$ dB errors are introduced by the equalizer, especially during the early part of the equalizer convergence interval as seen in

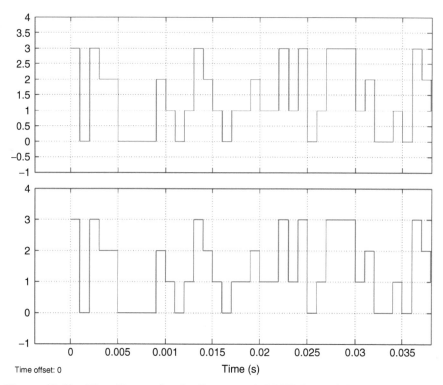

Figure 12.20 Time Traces for the Source and QPSK Demodulator with Digital Filter; (Top Trace is the Source and Bottom Trace is the QPSK Demodulator Output with Digital Filter and AWGN for $E_s/N_o = 50\,\text{dB}$).

Figure 12.21. As E_s/N_o decreases, the QPSK demodulated signal from the AWGN block and digital filter produces errors due to ISI. With equalization the resulting errors are reduced with symbol error probabilities that are close to the system without the digital filter. The data in Table 12.1 illustrate that ISI is significantly mitigated by the equalizer.

12.5 BER PERFORMANCE OF BPSK IN DISPERSIVE MULTIPATH CHANNEL USING A DECISION FEEDBACK EQUALIZER

Decision feedback equalization has been shown to have substantial improvement in diminishing the effects of multipath and/or ISI compared with performance achievable using linear LMS equalization. An implementation of a symbol-spaced DFE is shown in Figure 12.23, where it is observed that symbols leaving the detector are entered into a feedback transversal

DISPERSIVE MULTIPATH CHANNEL USING A DECISION FEEDBACK EQUALIZER 269

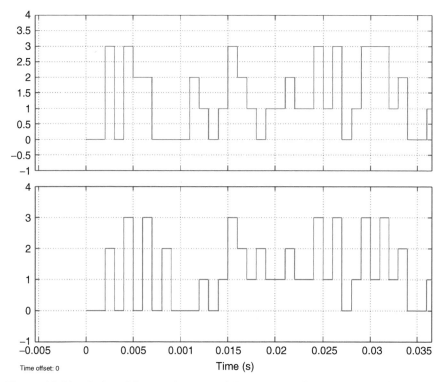

Figure 12.21 Delayed Source Output and QPSK Demodulator Output after Equalization; (Top Trace is Delayed Source Output; Bottom Trace is QPSK Demodulator Output after Equalization with $E_s/N_o = 50$ dB).

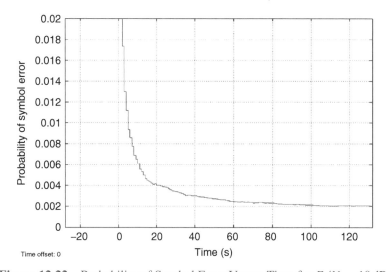

Figure 12.22 Probability of Symbol Error Versus Time for $E_s/N_o = 10$ dB.

TABLE 12.1 Probability of Symbol Error for Selected Values of E_s/N_o

E_s/N_o	Probability of Symbol Error (1000 s)		
	QPSK in AWGN	QPSK AWGN and Digital Filter	QPSK after Equalization
50	0	0	1.7×10^{-5}
15	0	0.00450	1.5×10^{-5}
10	0.0016	0.0416	0.0019
6	0.046	0.122	0.047

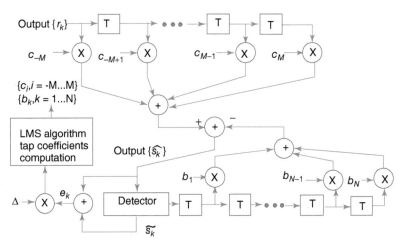

Figure 12.23 Decision Feedback Equalizer with $2M+1$ Feedforward and N Feedback Taps.

filter whose output is subtracted from the forward transversal filter output prior to detection.[6] In Figure 12.23, the DFE is seen to have $2M+1$ forward taps identified as $\{c_i, i = -M \ldots M\}$ and N feedback taps identified as $\{b_k, k = 1 \ldots N\}$. An error signal e_k, scaled by the factor Δ, is then used to compute both the feed forward and feedback taps via the LMS algorithm.

An example is now presented to compare the performance of a linear LMS equalizer with a decision feedback equalizer assuming BPSK modulation. The channel is a two-tap multipath channel spaced at the symbol rate with real gains (1, 1). Figure 12.24 depicts the Simulink simulation where $E_b/N_o = 3$ dB.

---

[6] As in the case of the linear LMS equalizer, delay elements in the forward filter may be fractionally spaced.

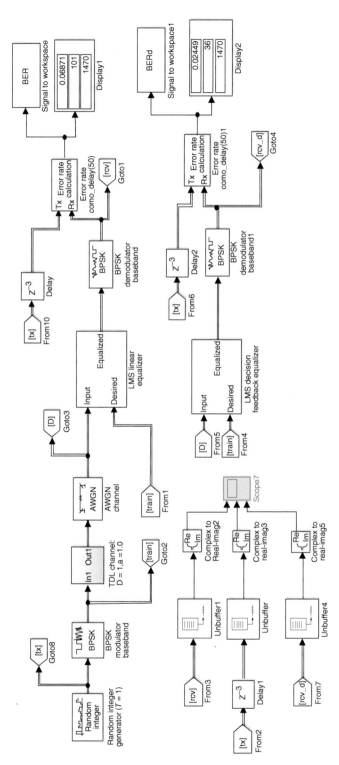

Figure 12.24 BER Performance of BPSK in a Multipath Channel Using Linear LMS Equalizer and a Decision Feedback Equalizer.

The Simulink model parameters for this example are specified as follows:

Model Parameters for Linear LMS and Decision Feedback Equalizer

- BPSK antipodal signals $= +1$ and -1 ($M = 2$)
- Symbol period $=$ sample time $= 1$ s
- Frame based with 10 samples/frame
- Simulation time $= 10,000$ s
- Input signal power $= 1$ W
- Two-path multipath with gains (1, 1) and symbol-spaced delay
- 6 tap LMS linear equalizer
- Decision feedback equalizer: 6 forward and 6 feedback taps
- Signal constellation for both equalizers $= (-1 \ 1)$
- Scale factor for both equalizers $\Delta = 0.01$
- For both equalizers synchronize received signal to tap 3 (reference tap)
- Number of samples/symbol $= 1$
- Leakage factor for both equalizers $= 1$
- Initial weights for both equalizers $= 0$
- Receive delay $= 0$ s; computation delay $= 50$ s
- Ideal equalizer training

In this model, both equalizers use tap 3 as the reference for synchronization.

Figure 12.25 displays the outputs of the source and both equalizers. The top trace is the BPSK demodulator output using the linear LMS equalizer; the middle trace is the source output delayed by three symbols; the bottom trace is the BPSK demodulator output using the decision feedback equalizer. A 50 s computation delay is used in both equalizers to allow them to converge before computing the BER. Agreement among the three traces is then observed after the 50 s computation delay.

Figure 12.26 depicts the BER performance for BPSK modulation in the equalized multipath channel. For comparison ideal BPSK BER performance is shown with no multipath. The results illustrate that in this severe multipath channel, DFE outperforms linear LMS equalization.

RECURSIVE LEAST SQUARES (RLS) EQUALIZER

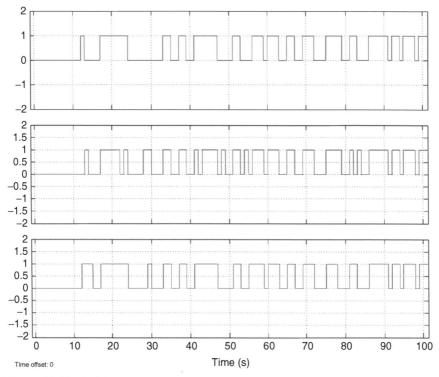

Figure 12.25 With $E_b/N_o = 3$ dB Top Trace BPSK Demodulator Output with Linear LMS Equalizer; Middle Trace Delayed Source Output; Bottom Trace BPSK Demodulator Output with Decision Feedback Equalizer.

12.6 BER PERFORMANCE OF BPSK IN RAYLEIGH FADING MULTIPATH CHANNEL USING AN RLS EQUALIZER

12.6.1 RLS Equalizer Description

RLS algorithms are known to have more rapid convergence than LMS algorithms albeit with greater complexity. As a consequence, RLS algorithms offer improved performance in time varying channels where fading is prevalent. A brief summary of the RLS algorithm is presented prior to proceeding with Simulink modeling.

RLS algorithm definitions are provided in Table 12.2.

The forgetting factor λ enables the RLS equalizer to reduce the influence of prior data on the estimated output. This factor then controls the rate of weight

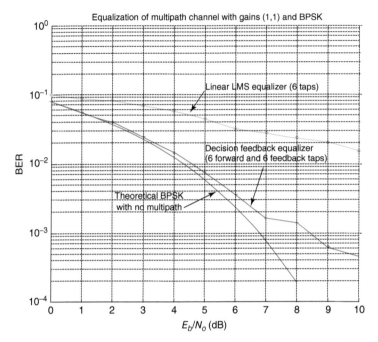

Figure 12.26 Comparison of Linear LMS and Decision Feedback Equalizer Performance.

convergence and enables rapid tracking of the channel's time variations. The RLS equations are presented as follows:

RLS estimated output[7]: $\hat{s}_k = \mathbf{w}_{k-1}^H \mathbf{r}_k$

A priori error: $e_k = d_k - \hat{s}_k$

Weighting coefficient recursion: $\hat{w}_k = \hat{w}_{k-1} + g_k e_k^*$

Gain computation: $g_k = \dfrac{\lambda^{-1} P_{k-1} r_k}{1 + \lambda^{-1} r_k^H P_{k-1} r_k}$

Inverse correlation matrix update: $P_k = \lambda^{-1}\{P_{k-1} - g_k r_k^H P_{k-1}\}$

The gain and correlation matrix expressions are similar to those occurring in Kalman filtering.[8]

---

[7] H denotes Hermitian transpose.
[8] Schonhoff, T.A., and A.A. Giordano, op.cit., Chapter 14.

RECURSIVE LEAST SQUARES (RLS) EQUALIZER

TABLE 12.2 RLS Algorithm Definitions

Variable	Description
N	RLS FIR filter order
$\widehat{w}_k$	Estimate of RLS, FIR filter weighting coefficients with $N+1$ taps ($k=0,1,\ldots,N$)
$\mathbf{w_k}$	Vector of weighting coefficient taps $\mathbf{w_k} = [\widehat{w}_0 \widehat{w}_1 \ldots \widehat{w}_N]$
r_k	Received signal
$\mathbf{r_k}$	Vector of current sample plus N most recent samples, $\mathbf{r_k} = [r_k\, r_{k-1}\, \ldots\, r_{k-N}]$
d_k	Desired signal
$\widehat{s}_k$	RLS estimated output, $\widehat{s}_k = \mathbf{w}_{k-1}^H \mathbf{r}_k$
e_k	Error signal, $e_k = d_k - \widehat{s}_k$ (a priori error before filter updates)
λ	Forgetting factor where $0 < \lambda < 1$
R_k	Weighted autocorrelation matrix of the received signal
P_k	Inverse of weighted autocorrelation matrix $P_k = R_k^{-1}$
g_k	Gain
δ	Initial value of P_k

12.6.2 RLS Equalization in Rayleigh Fading with No Multipath

A Simulink model used to illustrate RLS equalizer performance for a Rayleigh fading channel is shown in Figure 12.27. An LMS equalizer is included in the model in order to compare the performance of the RLS and LMS equalizers. With no multipath and a high $E_b/N_o = 10$ dB both equalizers converged within 15 s as seen in Figure 12.28; it is noted in this figure that proper synchronization is achieved. Since there is no computation delay, the first pulse in Figure 12.28, associated witht the source output, indicates that both equalizers experience errors during initial convergence. Figure 12.29 illustrates the Rayleigh fading channel magnitudes at 100, 1000, and 10,000 s, thus illustrating the time-varying behavior of the channel. For the selected model parameters, it is seen in Figure 12.30 that the RLS equalizer is able to track the channel fading whereas the linear LMS equalizer is not. As a benchmark, theoretical BER performance for BPSK in a Rayleigh fading channel is also depicted.

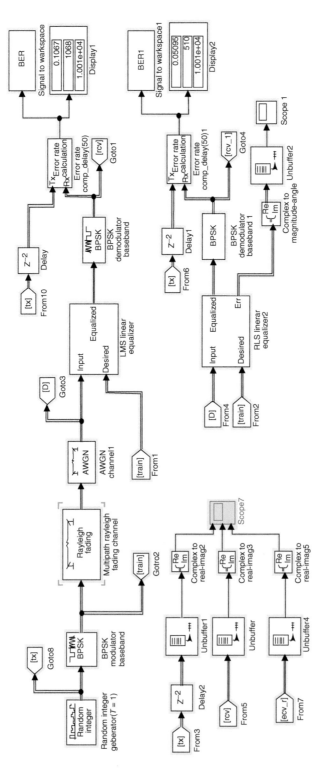

Figure 12.27 RLS and LMS Equalizers in Rayleigh Fading ($E_b/N_o = 10$ dB, Doppler $= 0.001$ Hz).

RECURSIVE LEAST SQUARES (RLS) EQUALIZER

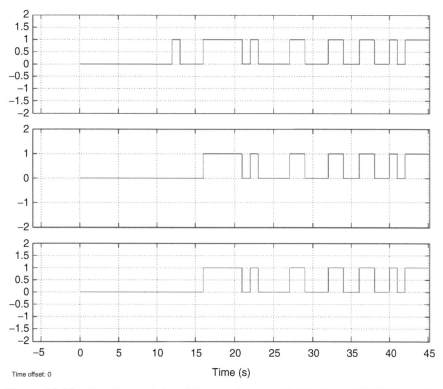

Figure 12.28 Top Trace: Delayed Source Output; Middle Trace: BPSK Demodulator Output with Linear LMS Equalizer; Bottom Trace: BPSK Demodulator Output with RLS Equalizer.

The Simulink model parameters for this example are specified as follows:

Model Parameters for Linear LMS and RLS Equalizer

- BPSK antipodal signals $= +1$ and -1 ($M = 2$)
- Symbol period = sample time = 1 s
- Frame based with 10 samples/frame
- Simulation time = 10,000 s
- Input signal power = 1 W
- Rayleigh channel Jakes fading model, no multipath
- Maximum Doppler shift = 0.001 Hz
- 4 tap LMS linear equalizer

- 4 tap RLS equalizer
- Signal constellation for both equalizers = (−1 1)
- Scale factor for LMS equalizer $\Delta = 0.01$
- For both equalizers synchronize received signal to tap 2 (reference tap)
- Number of samples/symbol = 1
- Linear LMS Leakage factor = 1
- RLS forgetting factor = 0.5
- RLS inverse correlation matrix = 0.1*eye(4)
- Initial weights for both equalizers = 0
- Receive delay = computation delay = 0 s
- Ideal equalizer training
- $E_b/N_o = 10$ dB

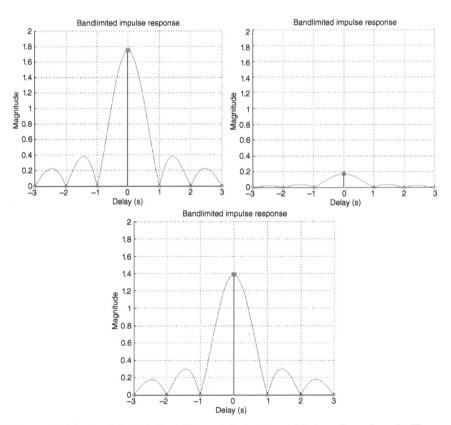

Figure 12.29 Rayleigh Fading Channel Behavior at Various Snapshots in Time. (Upper left = 100 s, upper right = 1,000 s and bottom = 10,000 s).

RECURSIVE LEAST SQUARES (RLS) EQUALIZER

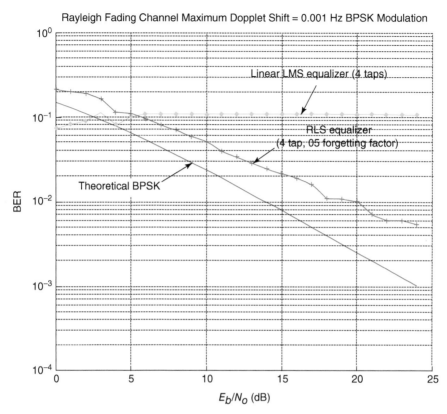

Figure 12.30 RLS and Linear LMS Equalizer BPSK BER Performance for a Rayleigh Fading Channel (Note: Stop BER Simulation with 200 Errors).

12.6.3 RLS Equalization in Rayleigh Fading with Multipath

Figure 12.31 presents a Simulink model to illustrate RLS equalizer performance for a Rayleigh fading channel with multipath. The channel includes two-path multipath with a delay vector = [0 1] s and multipath gains = [0 −3] dB. Figure 12.32 depicts RLS and linear LMS equalizer BER performance for BPSK in a Rayleigh fading channel with two-path multipath. Once again, the RLS equalizer exhibits better performance than the linear LMS equalizer.

The Simulink model parameters for this example are listed as follows:

Model Parameters for Linear LMS and RLS Equalizer

- BPSK antipodal signals = +1 and −1 ($M = 2$)

- Two-path channel with delay vector = [0 1] s and multipath gains = [0 –3] dB
- Symbol period = sample time = 1 s
- Frame based with 10 samples/frame
- Simulation time = 10,000 s
- Input signal power = 1 W
- Rayleigh fading channel Jakes model
- Rayleigh channel maximum Doppler shift = 0.001 Hz
- Signal constellation for both equalizers = (−1 1)
- Scale factor for linear LMS equalizer $\Delta = 0.01$
- 8 tap linear LMS equalizer
- 8 tap RLS equalizer
- For both equalizers synchronize received signal to tap 4 (reference tap)
- Number of samples/symbol = 1
- Linear LMS Leakage factor = 1
- RLS forgetting factor = 0.8
- RLS inverse correlation matrix = 0.1*eye(8)
- Initial weights for both equalizers = 0
- Receive delay = computation delay = 0 s
- Ideal equalizer training
- $E_b/N_o = 10$ dB

12.7 SUMMARY DISCUSSION

Equalization is an important technique for mitigating multipath and ISI. Equalizers considered here included linear LMS, LMS decision feedback, and RLS. The equalizer choice is a tradeoff between the rate of convergence and the implementation complexity. Decision feedback equalizers were shown to offer better performance than linear LMS equalizers. For fading channels with rapid time variation, an RLS equalizer offers the best performance. In all cases, final performance is dictated by the selection of relevant parameters such as the scale factor in LMS equalizers and the forgetting factor in RLS equalizers.

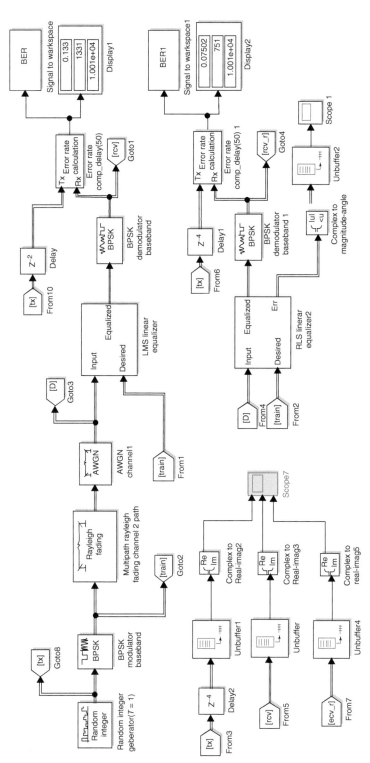

Figure 12.31 RLS and LMS Equalizers in Rayleigh Fading with Two-Path Multipath ($E_b/N_o = 10$ dB, Multipath delay vector = [0 1] and Multipath gains = [0 −3] dB).

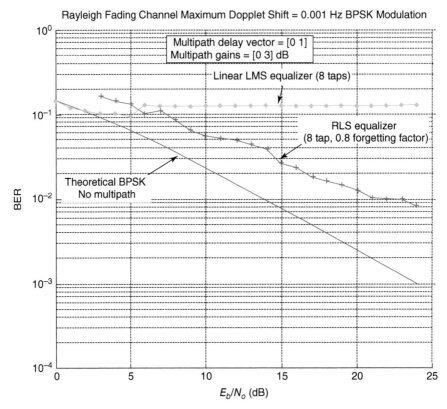

Figure 12.32 RLS and Linear LMS Equalizer BER Performance for BPSK in a Rayleigh Fading Channel with Multipath (Note: Stop BER Simulation with 200 Errors).

PROBLEMS

12.1 Change the scale factor in the Simulink model from 0.01 to (i) 0.1 and (ii) 0.001. Plot the error magnitude and the real part of the center tap coefficient as a function of time and compare the results to the case with the 0.01 scale factor.

12.2 Modify the Simulink model shown in Figures 12.4 to 12.6 to use a two-tap adaptive LMS equalizer with $p = 0$ for synchronization and coefficients c_0 and c_1.

1. Display the Simulink model as in Figure 12.4a with $E_b/N_o = 10$ dB
2. Identify the normal equations in matrix form for this case
3. Compute the ideal equalizer coefficients
4. Display the magnitude of the error and the real part of the two equalizer coefficients versus time for a simulation of 10,000 s
5. Using bertool plot the BER with and without the equalizer along with the theoretical result with no multipath

12.3 Modify the Simulink model in Figure 12.24 to implement a channel with multipath gains (1 0.5). Using the bertool obtain the BER for the linear LMS and decision feedback equalizers and compare the results to the channel with no multipath.

12.4 Construct a Simulink model to compare the performance of the RLS decision feedback equalizer from the Simulink library with the linear RLS equalizer assuming BPSK modulation over a two-path multipath channel with multipath gains = [0 –3] dB and delay vector = [0 1] s.

Assume that the RLS decision feedback equalizer has 8 feed forward and 8 feedback taps with a 0.9 forgetting factor. Assume that the linear RLS equalizer has 8 taps with 0.8 forgetting factor. Use 10 samples/frame, 1 s symbol time and 1 sample/s.

1. Execute the Simulink model for 10,000 s with $E_b/N_o = 15$ dB and display the model.
2. List the Simulink model parameters.
3. Plot the source output, the RLS linear equalized BPSK demodulator output, and the RLS decision feedback equalized BPSK demodulator output to demonstrate that proper synchronization has been achieved.

4. Plot the theoretical BPSK BER with no multipath, the BER from the BPSK demodulator output with linear RLS equalization, and the BER from the BPSK demodulator output with decision feedback RLS equalization. Assume 2 dB steps and use a range of E_b/N_o values between 0 and 24 dB.

13

SIMULINK EXAMPLES

Simulink modeling is employed in many areas of engineering, including digital communications systems, which is the focus of this book. This chapter presents a number of illustrative examples; the topics covered here include the following:

- Linear predictive coding (LPC) for speech compression
- RLS interference cancellation
  - Sinusoidal interference
  - Low pass filtered Gaussian noise
- Spread spectrum
  - Sinusoidal interference
  - Excision using an RLS canceller
- Antenna nulling of a single interferer
- Kalman filtering
  - Scalar Kalman
  - Kalman equalizer
  - Radar tracking using extended Kalman filter (EKF)

Modeling of Digital Communication Systems Using SIMULINK®, First Edition.
Arthur A. Giordano and Allen H. Levesque.
© 2015 John Wiley & Sons, Inc. Published 2015 by John Wiley & Sons, Inc.
Companion Website: www.wiley.com/go/simulink

- Orthogonal frequency division multiplexing (OFDM)
- Turbo coding with BPSK

The models described here are available on the companion website, www.wiley.com/go/simulink. The reader is encouraged to execute the models and experiment with the parameters of the model and selections made for running the model.

13.1 LINEAR PREDICTIVE CODING (LPC) FOR SPEECH COMPRESSION

LPC is a well-known technique for compressing speech. It is based on an all-pole model where a current speech sample $s(n)$ is estimated by a linear sum of L prior samples weighted with prediction coefficients a_k, $k = 1 \ldots L$, thus producing an error signal $e(n)$, that is,

$$e(n) = s(n) - \sum_{k=1}^{L} a_k s(n-k)$$

The sum is an estimate of the speech segment and is denoted by

$$\hat{s}(n) = \sum_{k=1}^{L} a_k s(n-k)$$

The physical basis of this model is that the vocal tract can be represented by a combination of voiced and unvoiced sounds as well as silent intervals. For voiced sounds, the vocal cords vibrate, opening and closing at a rate corresponding to the pitch of the speaker's voice. Unvoiced sounds, known as plosive, due to mouth closure, and fricative, due to narrow passage of air through the mouth, occur in the case where the vocal cords do not vibrate. Figure 13.1 depicts a mathematical model of the vocal tract.

The pitch period, P, for human speech typically ranges from 2.5 to 10 ms. The gain G, representing the volume of air passing through the vocal tract, determines the loudness of the voice. The all-pole filter corresponding to the LPC model is given by the transfer function $H(z)$

$$H(z) = \frac{1}{1 + a_1 z^{-1} + a_2 z^{-2} \cdots a_L z^{-L}}$$

The coefficients are determined by minimizing the mean square error (MSE) and lead to the normal equations expressed in terms of correlations

LINEAR PREDICTIVE CODING (LPC) FOR SPEECH COMPRESSION

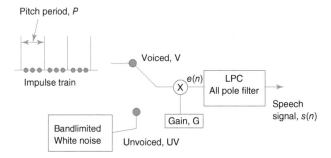

Figure 13.1 Speech Model of the Vocal Tract.

of the speech signal. Defining the MSE over a short segment of speech as

$$E = \sum_n e^2(n) = \sum_n (s^2(n) - \hat{s}^2(n))$$

The speech signal speech correlations can be expressed as

$$R(k) = \sum_n s(n)s(n+k)$$

The normal equations are then given by

$$\begin{bmatrix} R(0) & \cdots & R(L-1) \\ \cdots & \vdots & \cdots \\ R(L-1) & \cdots & R(0) \end{bmatrix} \begin{bmatrix} a_1 \\ \vdots \\ a_L \end{bmatrix} = \begin{bmatrix} R(1) \\ \vdots \\ R(L) \end{bmatrix}$$

The prediction coefficients in the normal equations are computed efficiently by applying the Levinson Durbin algorithm to the $L \times L$ Toeplitz matrix.[1]

During speech compression, an all-pole analysis filter is used to estimate the prediction coefficients. These coefficients are then sent to the receiver where an all-pole synthesis filter is used to reconstruct the speech. Thus instead of sending the speech signal itself at a high transmission rate, a lower transmission rate is used to transmit the prediction coefficients instead.

Figure 13.2 depicts a Simulink LPC speech model comprising three elements: (i) speech vocal tract model, (ii) computation of prediction coefficients, and (iii) speech analysis and synthesis.

This Simulink model uses a variable-step with a discrete solver. The samples are generated at intervals of 0.0001 s.

---

[1] Giordano, A.A., and F.H. Hsu, *Least Square Estimation with Application to Digital Signal processing*, Wiley, 1985.

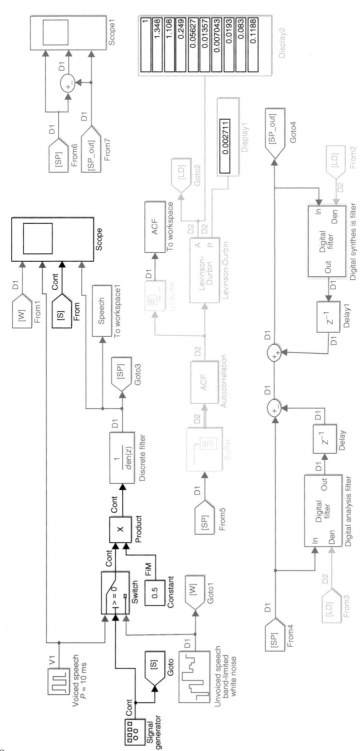

Figure 13.2 LPC Speech Model with Analysis and Synthesis.

13.1.1 Speech Vocal Tract Model

A 100 Hz signal generator switches between voiced and unvoiced signals. The voiced signal consists of a train of impulses with a pitch period of 10 ms. The unvoiced signal is bandlimited white noise. The gain is arbitrarily set to be 0.5. The multiplier output is passed through an all-pole filter with its six denominator coefficients selected to be [1 0.1 0.2 −0.3 0.4 −0.05]. The output of this filter then represents a model of a speech signal.

13.1.2 Prediction Coefficients Computation

The speech signal is buffered with a block size of 10 in order to generate 10 prediction coefficients. The correlations are computed using the autocorrelation function (ACF) block with input from the buffer. The ACF output is then passed to the Levinson_Durbin block to determine the 10 prediction coefficients. The 10 coefficients are displayed along with the prediction error power.

13.1.3 Speech Analysis and Synthesis

The predictor coefficients are then delivered to both the analysis and synthesis filters. The analysis filter accepts the speech input and produces the error signal. The synthesis filter accepts the error signal and reconstructs the speech signal.

The Simulink model parameters for this example are listed as follows:

Model Parameters for LPC Speech Model

- Voiced speech: 10 ms period, 1% pulse width and amplitude 1.0
- Unvoiced speech: bandlimited white noise: 5 mW, 100 μs sample time
- Switch voice to unvoiced every 10 ms
- Vocal tract all pole filter with coefficients [1 0.1 0.2 −0.3 0.4 −0.05]
- Speech model gain: 0.5
- 10 LPC predictor coefficients computed by Levinson_Durbin
- All pole analysis and synthesis digital filters
- No additive noise

Figure 13.3 depicts the signals used to generate the model of the speech. From top to bottom, the traces are: the bandlimited white noise, the impulse train with a 10 ms period, the 100 Hz signal generator, and the speech signal. Note that the speech signal exhibits bursts with silent periods. Figure 13.4

290 SIMULINK EXAMPLES

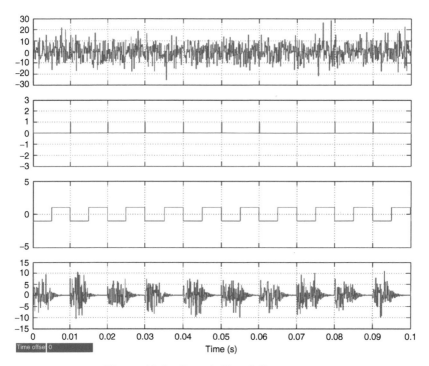

Figure 13.3 Speech Signal Components.

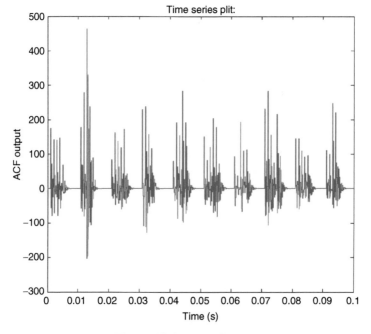

Figure 13.4 ACF Output.

RLS INTERFERENCE CANCELLATION

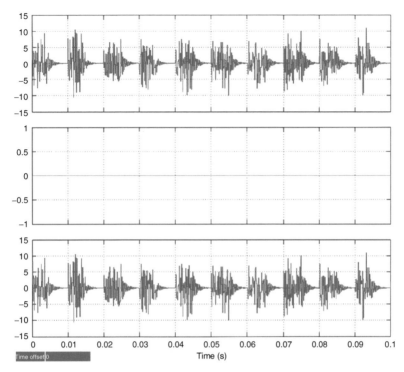

Figure 13.5 Speech Signal; Error signal; Reconstructed Speech Signal (Ideal Case).

illustrates the output of the ACF block where the pitch period is observed to be about 10 ms.

Figure 13.5 illustrates the original speech signal in the top trace, the reconstructed speech signal in the bottom trace, and the error signal in the middle trace. Since this simulation is idealized, the error signal is zero.

13.2 RLS INTERFERENCE CANCELLATION

This section explores Simulink models that accomplish interference mitigation using the RLS algorithm. BPSK modulation is selected, where two demodulators are used in order to compare the BER (bit error rate) performance with and without the canceller.

13.2.1 Sinusoidal Interference

Figure 13.6 depicts the Simulink model where two sinusoids are summed and amplified to produce the interference. The RLS filter has length 8 with outputs that include the error signal and filter weights in addition to the filtered output.

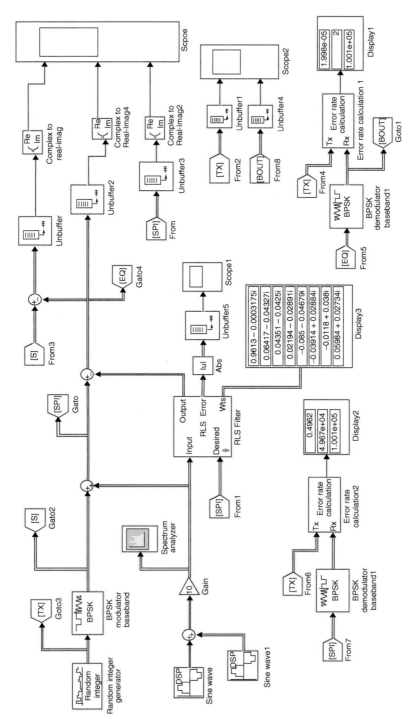

Figure 13.6 RLS interference Canceller.

The Simulink model parameters for this example are listed as follows:

Model Parameters for Interference Canceller with Sinusoidal Interference

- Frame-based simulation with 100 samples/frame
- Simulation time: 100 s
- Sample time: 0.001 s
- BPSK modulation
- Two sinusoids with 100 and 300 Hz frequencies summed and amplified by 10
- RLS filter length = 8, forgetting factor = 1, initial input variance = 0.1
- No additive noise

The 100 s simulation shown in Figure 13.6 yields a BER = 2×10^{-5} when the RLS filter is employed and ~0.5 BER when no RLS filter is used. The eight complex filter weights are seen in Figure 13.6. The power spectrum for the two amplified sinusoids is displayed in Figure 13.7.

In the RLS filter, the desired signal is the signal plus interference and the input is the interference alone. The RLS filter output is then an estimate of the interference, which is then subtracted from the signal-plus-interference to form an estimate of the transmitted signal labeled as EQ in Figure 13.6.

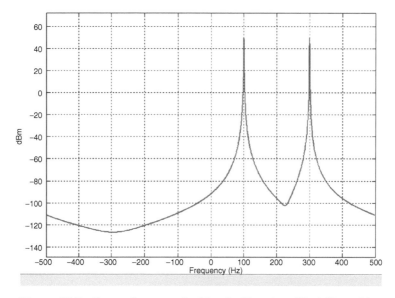

Figure 13.7 Power Spectrum in dBm for Two Amplified Sinusoids.

The estimated transmitted signal is sent to the BPSK demodulator to produce an output that over this simulation run produces only two errors, which occur while the RLS algorithm is converging. The signal-plus-interference labeled SPI in Figure 13.6 does not have the benefit of the canceller and thus produces all errors. Figure 13.8 displays a partial output of the real part of three signals from the scope: (i) the top trace is the difference between the BPSK modulator output and the RLS filter output, (ii) the middle trace is the RLS filter output and the bottom trace is the signal plus interference. It is evident from Figure 13.8 that the difference signal in the top trace is converging toward zero as the RLS filter converges.

Figure 13.9 displays the source output (top trace) and BPSK demodulator output (bottom trace), where the RLS filter has cancelled the interference. This figure illustrates that the two errors occur early as the RLS filter converges and no errors occur thereafter.

Figure 13.10 illustrates the rapid convergence of the error magnitude observed in the RLS filter. Note that in Figures 13.7 to 13.10, a 0.1 s delay occurs due to the 0.1 s frame length.

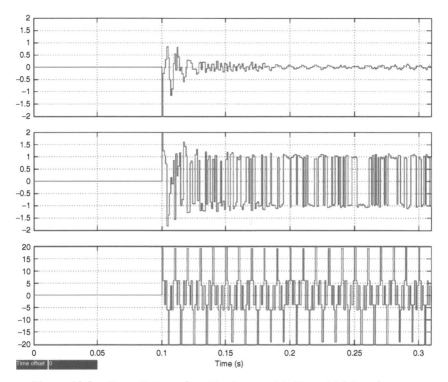

Figure 13.8 Three Outputs from the Scope with Sinusoidal Interference.

RLS INTERFERENCE CANCELLATION

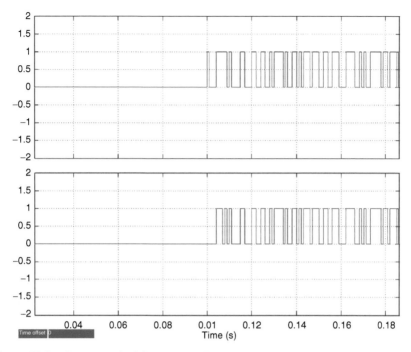

Figure 13.9 Source and BPSK Demodulator Outputs with Sinusoidal Interference.

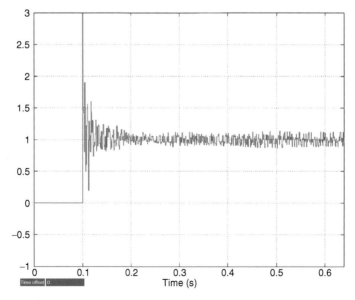

Figure 13.10 RLS Filter Error Magnitude with Sinusoidal Interference.

TABLE 13.1 BER Results with and without the RLS Canceller for Sinusoidal Interference

Gain	RLS BPSK Demodulator BER	BPSK Demodulator BER without RLS Filter
0.5	10^{-5}	0.049
1.0	2×10^{-5}	0.099
3.0	2×10^{-5}	0.496
10.0	2×10^{-5}	0.496
30.0	2×10^{-5}	0.496

Table 13.1 presents the BER results for the two demodulators with different gain values.

From Table 13.1, it is observed that low values of gain cause the demodulator without an RLS filter to make fewer errors. In all cases, the BPSK demodulator with an RLS filter effectively cancels the interference.

13.2.2 Low Pass Filtered Gaussian Noise

Figure 13.11 displays the Simulink model for an RLS canceller where the interference is added as low pass filtered complex Gaussian noise. The 100 s simulation shown in Figure 13.11 yields a BER = 1.1×10^{-4} when the RLS filter is employed and ~0.44 BER when no RLS filter is used.

The Simulink model parameters for this example are listed as follows:

Model Parameters for Interference Canceller with Low Pass Filtered Gaussian Noise

- Sample-based simulation
- Simulation time = 100 s
- Sample time = 0.001 s
- BPSK modulation
- Complex Gaussian noise with unity variance in the real and imaginary parts
- Low pass filter: Equiripple finite duration impulse response (FIR) design, fpass = 0.45, fstop = 0.55, Apass = 1, and Astop = 60
- RLS filter length = 8, forgetting factor = 1, and initial input variance = 0.1
- No additive noise

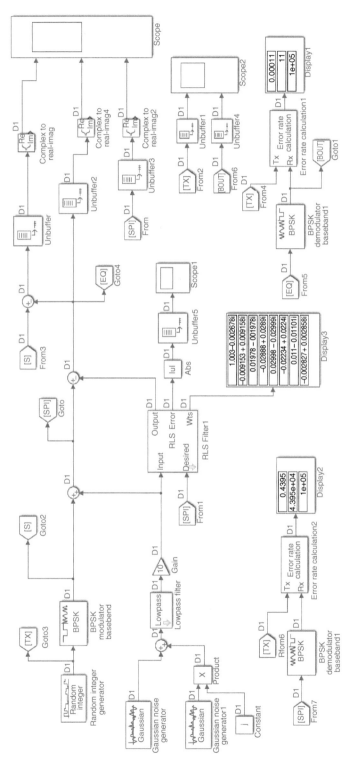

Figure 13.11 RLS Interference Canceller with Low Pass Filtered Gaussian Noise.

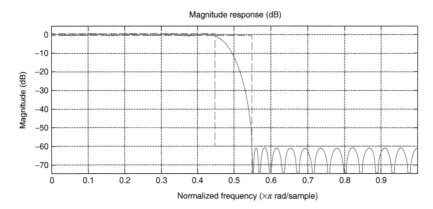

Figure 13.12 Low Pass Filter Frequency Response.

Figure 13.12 illustrates the frequency response of the low pass filter.

Figure 13.13 displays a partial output of the real part of the three outputs of the scope: (i) top trace is the difference between the BPSK modulator output and the RLS filter output, (ii) the middle trace is the RLS filter output, and (iii) the bottom trace is the signal plus interference.

Figure 13.14 displays the partial source output (top trace) and BPSK demodulator output (bottom trace), where the RLS filter has cancelled the interference. This figure illustrates that errors occur early as the RLS filter converges and no errors occur thereafter.

Figure 13.15 illustrates the rapid convergence of the error magnitude observed in the RLS filter.

Table 13.2 presents the BER results for the two demodulators with different gain values.

Once again Table 13.2 illustrates that low values of gain cause the demodulator without an RLS filter to make fewer errors. In all cases, the BPSK demodulator with an RLS filter effectively cancels the interference.

13.3 SPREAD SPECTRUM

13.3.1 Spread Spectrum Simulink Model without In-Band Interference

Spread spectrum communications offers a significant benefit in performance in the presence of in-band interference and/or jamming. Figure 13.16 depicts a spread spectrum system using BPSK modulation for both the data

SPREAD SPECTRUM

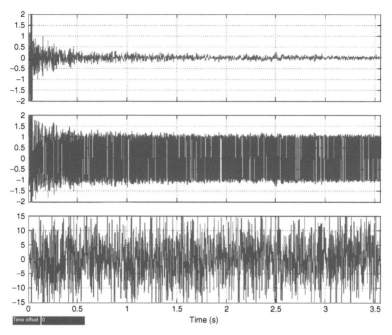

Figure 13.13 Three Outputs from the Scope with Gaussian Noise.

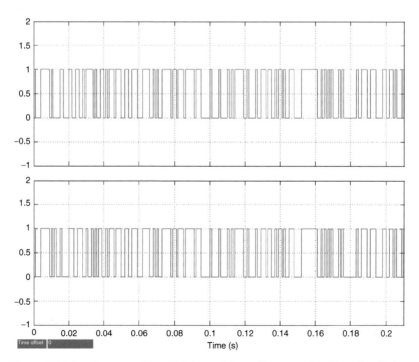

Figure 13.14 Source and BPSK Demodulator Outputs with Gaussian Noise.

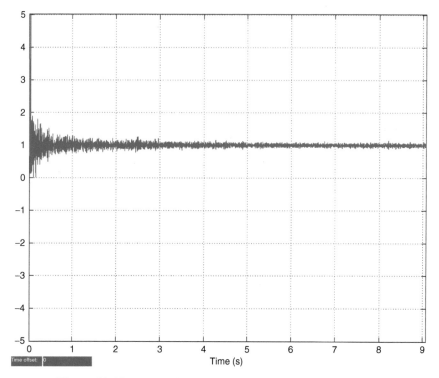

Figure 13.15 RLS Filter Error Magnitude with Gauss Noise.

TABLE 13.2 BER Results with and without the RLS Canceller for Gaussian Noise

Gain	RLS BPSK Demodulator BER	BPSK Demodulator BER without RLS Filter
0.5	5×10^{-5}	0.00196
1.0	10^{-4}	0.071
3.0	9×10^{-5}	0.3111
10.0	1.1×10^{-4}	0.4395
30.0	1.5×10^{-4}	0.4775

symbols and the spreading code. In this example, the spreading code has 10 chips/bit resulting in a 10dB process gain. Additive noise is included but no interference is added, in order to illustrate the spread spectrum system characteristics.

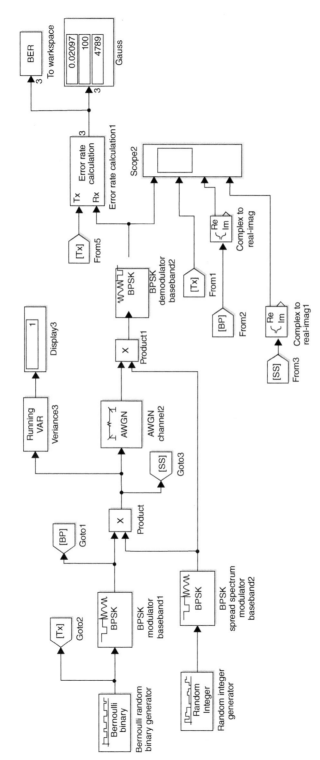

Figure 13.16 Spread Spectrum System with BPSK Modulation for Data and Spreading; AWGN without Interference.

The Simulink model parameters for this example are listed as follows:

Model Parameters for Spread Spectrum System without Interference

- Sample-based simulation
- Simulation time = 100 s
- Data symbol sample time = 1 s
- Chip symbol sample time = 0.1 s
- BPSK modulation for data and spreading
- Data source seed = 25741
- Spreading code source seed = 37
- $E_b/N_o = 3$ dB => BER = 0.02097
- Stop simulation with 100 errors.

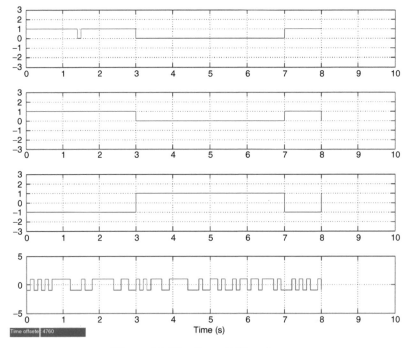

Figure 13.17 Scope2 Waveforms.

Figure 13.17 illustrates the following outputs from scope2. The traces from top to bottom are:

- Top = BPSK demodulator output

- Second one down = data source output
- Third one down = BPSK data modulator output
- Bottom = spread spectrum output

Here, to illustrate the effect of the spreading code on the modulated data, only 10 chips/bit are used. In Figure 13.17, the effect of spreading is observed most easily by examining the interval between 7 and 8 s.

The simulation gives BER = 0.021 for 100 errors. The theoretical BER for 3 dB E_b/N_o is

$$P_b = \frac{1}{2}\text{erfc}\left(\sqrt{\gamma_b}\right) = \frac{1}{2}\text{erfc}\left(\sqrt{10^{0.3}}\right) = 0.0229$$

The results demonstrate that spreading has no effect on BER performance in this case.

13.3.2 Spread Spectrum Simulink Model with In-Band Interference

It is now instructive to include in-band interference. For simplicity, an interfering sinusoidal signal is introduced as seen in the Simulink model shown in Figure 13.18.

The Simulink model parameters for this example are listed as follows:

Model Parameters for Spread Spectrum System with Interference

- Sample-based simulation
- Simulation time = 100 s
- Data symbol sample time = 1 s
- Chip symbol sample time = 0.01 s
- BPSK modulation for data and spreading
- Data source seed = 25741
- Spreading code source seed = 37
- No additive noise
- Sinusoid parameters: frequency = 1 kHz, amplitude = 0.5, sample time = 0.0001 s
- Stop simulation at 100 s

Figure 13.19 depicts the spectrum of the 1 kHz complex sinusoid as observed in Spectrum Analyzer2. In Figure 13.20, Spectrum Analyzer1 shows the spectrum of the spread spectrum waveform with 100 chips/bit.

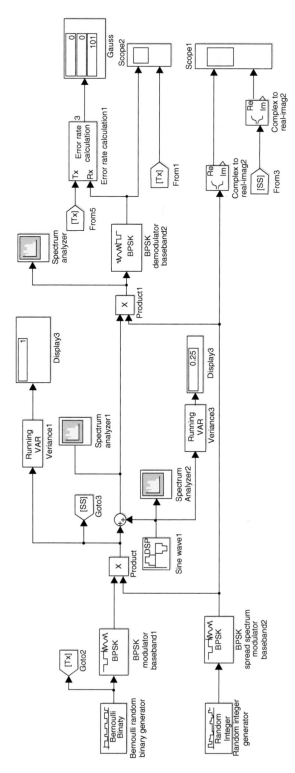

Figure 13.18 Spread Spectrum System with BPSK Modulation for Data & Spreading; with Interference and no Additive Noise (sinusoid amplitude = 0.5).

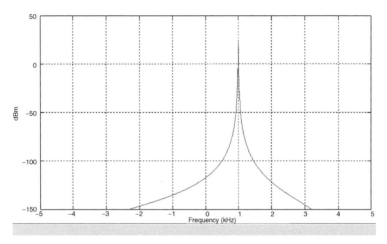

Figure 13.19 1 kHz Sinusoid in Spectrum Analyzer2 (sinusoid amplitude = 0.5).

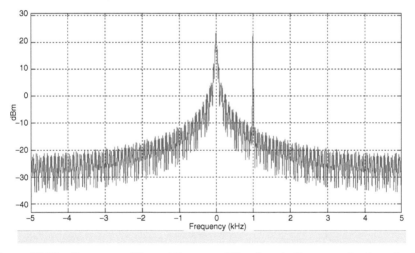

Figure 13.20 Spectrum of Spread Spectrum Waveform in Spectrum Analyzer1 with 100 Chips/Bit (sinusoid amplitude = 0.5).

The peak for the 1 kHz complex sinusoid is visible along with the data observed at 0 kHz. As shown in Figure 13.21, increasing the spreading to 1000 chips/bit raises the background level. Figure 13.22 depicts the spectrum of the despread waveform at the input to the BPSK data demodulator. Figure 13.23 shows the Scope1 results where the top and bottom traces show the outputs of the BPSK spread spectrum modulator and the spread spectrum waveform, respectively. Figure 13.24 shows that the BPSK data demodulator

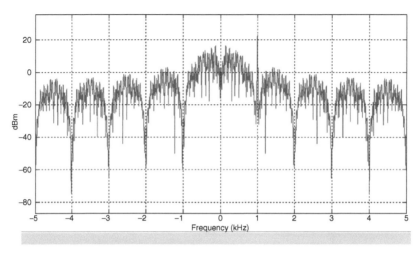

Figure 13.21 Spectrum of Spread Spectrum Waveform in Spectrum Analyzer1 with 1000 Chips/Bit, (sinusoid amplitude = 0.5).

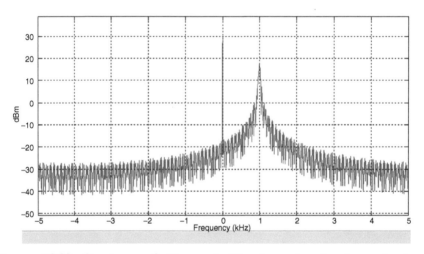

Figure 13.22 Spectrum of Despread Waveform in Spectrum Analyzer (sinusoid amplitude = 0.5).

output and source data output are in complete agreement, that is, no errors are made with a sinusoidal interference that has such a small amplitude.

As shown in Figure 13.25, the Simulink spread spectrum model with interference and 100 chips/bit is re-executed where the only change is to select an amplitude of 30 for the sinusoid. The output from the error rate calculation demonstrates that errors are now observed. Figure 13.26 depicts the spectrum of the 1 kHz sinusoid with an amplitude of 30.

SPREAD SPECTRUM

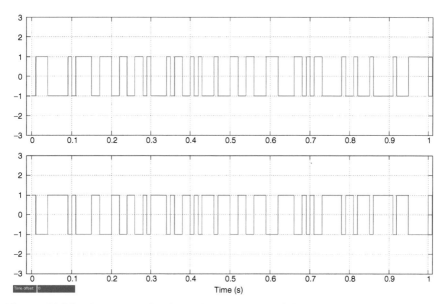

Figure 13.23 Scope1: BPSK Spread Spectrum Modulator (Top) and Spread Spectrum Waveform (Bottom) (sinusoid amplitude = 0.5).

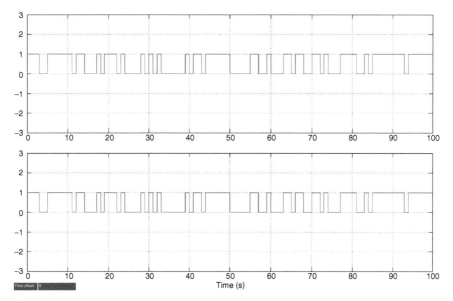

Figure 13.24 Scope2: BPSK Data Demodulator Output (Top); Source Data Output (Bottom) (sinusoid amplitude = 0.5).

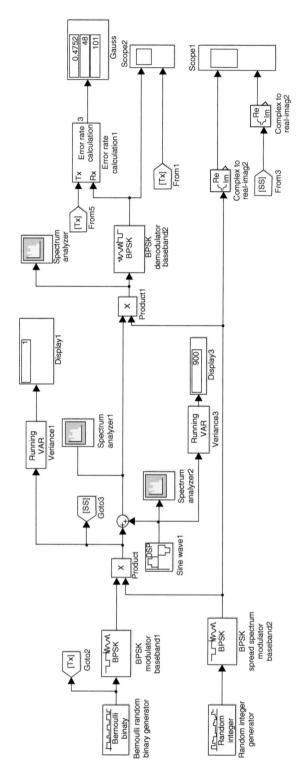

Figure 13.25 Spread Spectrum System with BPSK Modulation for Data and 100 Chips/Bit Spreading with Interference and no Additive Noise (sinusoid amplitude = 30).

SPREAD SPECTRUM 309

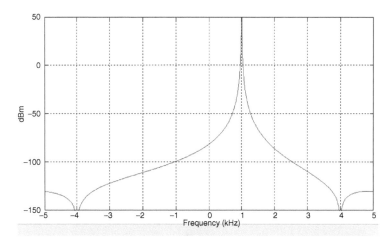

Figure 13.26 1 kHz Sinusoid in Spectrum Analyzer2 (sinusoid amplitude = 30).

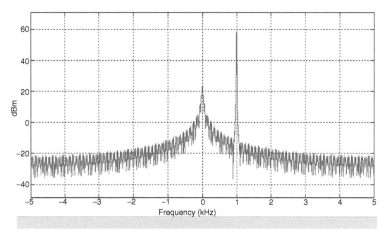

Figure 13.27 Spectrum of Spread Spectrum Waveform in Spectrum Analyzer1 with 100 Chips/Bit (sinusoid amplitude = 30).

Figures 13.27 and 13.28 show, respectively, the spectrum of the spread spectrum waveform and the despread waveform. As shown in Figure 13.28, the large peak at 1 kHz accounts for the errors in this simulation.

13.3.3 Spread Spectrum Simulink Model with In-Band Interference and Excision

It is well known that use of excision to suppress narrowband interference in a spread spectrum system substantially improves performance. The simulation

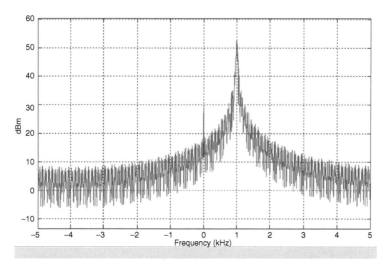

Figure 13.28 Spectrum of Despread Waveform in Spectrum Analyzer (sinusoid amplitude = 30).

provided in Figure 13.29 uses an RLS filter to cancel the interference. As a result, this simulation demonstrates that the interference is effectively eliminated and no errors occur at the demodulator output.

The Simulink model parameters for this example are listed as follows:

Model Parameters for Spread Spectrum System with Interference and RLS Cancellation

- Sample-based simulation
- Simulation time = 100 s
- Data symbol sample time = 1 s
- Chip symbol sample time = 0.01 s
- BPSK modulation for data and spreading
- Data source seed = 25741
- Spreading code source seed = 37
- No additive noise
- Sinusoid parameters: frequency = 1 kHz, amplitude = 30, sample time = 0.0001 s
- RLS filter parameters: filter length = 32, forgetting factor = 1, initial input variance = 0.1
- Stop simulation at 100 s

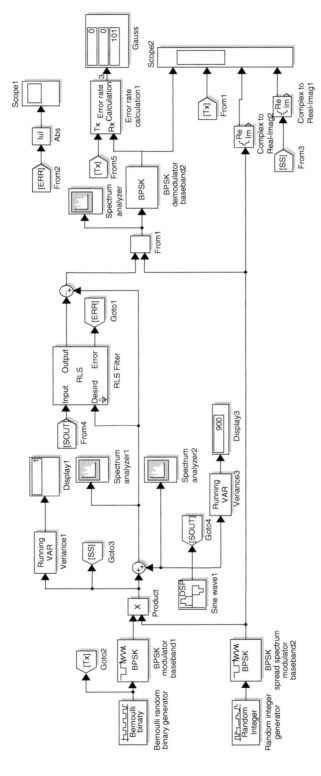

Figure 13.29 Spread Spectrum System with BPSK Modulation for Data & 100 Chips/Bit Spreading with Interference and RLS Canceller (No Additive Noise, sinusoid amplitude = 30).

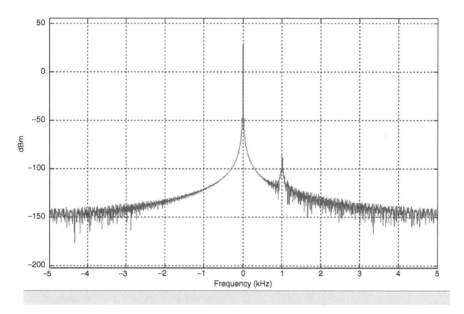

Figure 13.30 Despread Waveform in Spectrum Analyzer (sinusoid amplitude = 30).

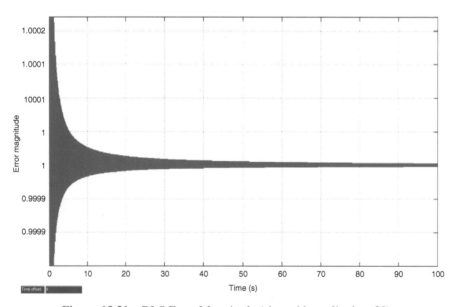

Figure 13.31 RLS Error Magnitude (sinusoid amplitude = 30).

Figure 13.30 showing the spectrum of the despread waveform indicates that the interference is greatly diminished. The RLS error magnitude shown in Figure 13.31 indicates that the RLS filter quickly reaches steady state.

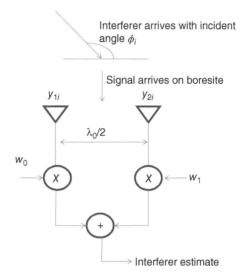

Figure 13.32 Antenna Nulling Model.

13.4 ANTENNA NULLING

Antenna nulling is a well-established technique for suppressing in-band interference from an arbitrary direction of arrival. In general, an N-element phased array can cancel $N-1$ interferers. A simple Simulink antenna nulling example will now be presented after a brief discussion of the theory.

Figure 13.32 depicts a two-element antenna array designed to estimate and cancel a single interferer. The array elements are separated by a half-wavelength $\lambda_o/2$. The signal is assumed to arrive on boresite whereas the interferer arrives at an incident angle ϕ_i. The array uses weights w_0 and w_1 to multiply the received signals y_{1i} and y_{2i}, respectively, on the individual branches. The output of the array is the estimate of the interferer.

The received signals are now represented in terms of the transmitted signal S_i and interferer I_i by

$$y_{1i} = S_i + I_i$$

$$y_{2i} = S_i + I_i e^{-j\beta r}$$

where for a wavelength λ, the phase factor $\beta = 2\pi/\lambda$ and $r = (\lambda_o/2)\cos\phi_i$. The product βr can be expressed in terms of frequencies using $f_o = c\lambda_o$ and $f = c/\lambda$

as $(\pi f \cos \phi_i)/f_o$ where c is the velocity of light, equal to 3×10^8 m/s. As an example for a GSM system operating at 900 MHz and an incident angle of 45°, $\beta r = 3\pi/\sqrt{2}$. To simplify the notation, let $\alpha = e^{-j\beta r}$ so that the second received signal is expressed as

$$y_{2i} = S_i + I_i\alpha$$

The estimated interferer is the weighted combination of the received signals and can be expressed as

$$\widehat{I}_i = w_0 y_{1i} + w_1 y_{2i}$$

The error signal is given by $e_i = I_i - \widehat{I}_i$.

Interference cancellation, producing an estimate of the desired signal, is accomplished prior to demodulation by forming the demodulator input as $y_{1i} - \widehat{I}_i$. To determine the weights, the orthogonality principle can be applied leading to the conditions

$$E\{y_{1i}^* e_i\} = 0 \quad \text{and} \quad E\{y_{2i}^* e_i\} = 0$$

From the aforementioned equations

$$E\{y_{1i}^* I_i\} = w_0 \ E\{y_{1i}^* y_{1i}\} + w_1 \ E\{y_{1i}^* y_{2i}\}$$

$$E\{y_{2i}^* I_i\} = w_0 \ E\{y_{2i}^* y_{1i}\} + w_1 \ E\{y_{2i}^* y_{2i}\}$$

Evaluating the moments with the assumptions that the signal and interference terms are orthogonal results in

$$1 = w_0(\gamma^{-1} + 1) + w_1(\gamma^{-1} + \alpha)$$

$$1 = w_0(\gamma^{-1}\alpha + 1) + w_1(\gamma^{-1} + 1)\alpha$$

where the interference to signal power ratio, $\gamma = E\{|I_i|^2\}/E\{|S_i|^2\}$. The simultaneous solution of these equations yields the weights $w_0 = 1/(1-\alpha)$ and $w_1 = -1/(1-\alpha)$. In the special case where $\gamma = 1$ and $\beta r = 3\pi/\sqrt{2}$ (and $\alpha = e^{-j\beta r}$), the weights are computed as $w_0 = 0.5 - j\,2.5919$, and $w_1 = -0.5 + j\,2.5919$.

A Simulink model for a noise-free example of interference nulling using an LMS algorithm is presented in Figure 13.33.

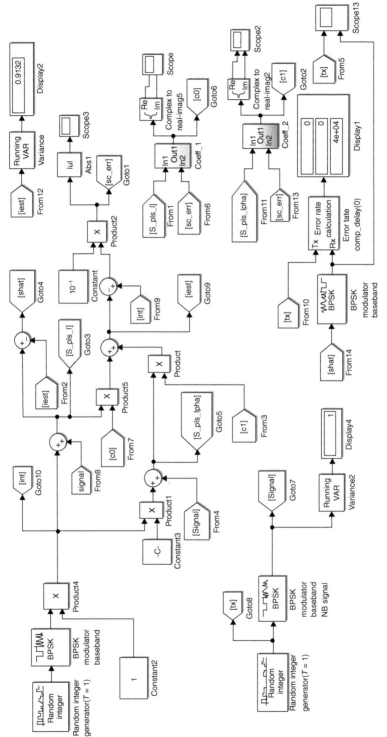

Figure 13.33 Noise Free Simulink Model of Interference Nulling, $S/I = 0$ dB.

The Simulink model parameters for this example are listed as follows:

Model Parameters for Antenna Nulling System with Interference and LMS Cancellation

- Sample-based simulation
- Simulation time = 40,000 s
- Data symbol sample time = 1 s
- Interference symbol sample time = 1 s
- Interference incident angle $\phi_i = 45°$
- Interference constant $C = \exp(-j3\pi\cos(\pi/4))$
- Interference amplitude = 1 V, $S/I = 0$ dB
- BPSK modulation for data and interference
- Data source seed = 37
- Interference source seed = 23
- No additive noise
- Two tap LMS algorithm with scale factor = 0.1

For clarity in this noise-free case, the signal to interference power ratio, $S/I = 0$ dB and $\beta r = \frac{2\pi}{\lambda} \frac{\lambda_0}{2} \cos\phi_i = \frac{\pi f}{f_0} \cos\phi_i$; for GSM at 900 MHz, $\frac{f}{f_0} = \frac{900 \times 10^6}{3 \times 10^8} = 3 = \beta r = 3\pi \cos\phi_i$.

Figure 13.34 shows the plots of the error magnitude obtained from scope3 depicted in Figure 13.33.

The real and imaginary parts of coefficients w_0 and w_1 are displayed in Figures 13.35 and 13.36, respectively. For this special case, it is observed that the simulated values of w_0 and w_1 are converging to the theoretical values computed earlier.

Figure 13.37 displays a comparison between the random source output and the demodulator output obtained from scope13 depicted in Figure 13.33, where the interference is seen to be effectively nulled over the 100 s observation interval. The error rate calculation shown in Figure 13.33 indicates that no errors occur over the 40,000 s simulation period.

Another Simulink simulation is presented in Figure 13.38, where the only change is to increase the interference power so that $S/I = -10$ dB.

ANTENNA NULLING

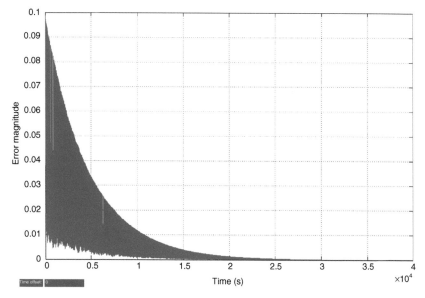

Figure 13.34 LMS Error Magnitude of Interference Nulling Example.

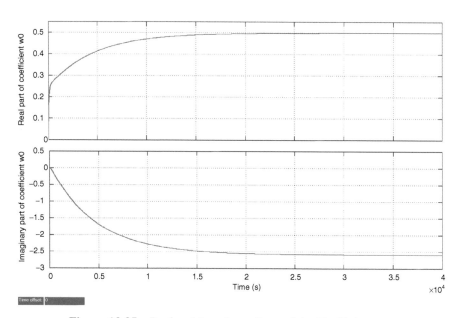

Figure 13.35 Real and Imaginary Parts of the Coefficient w_0.

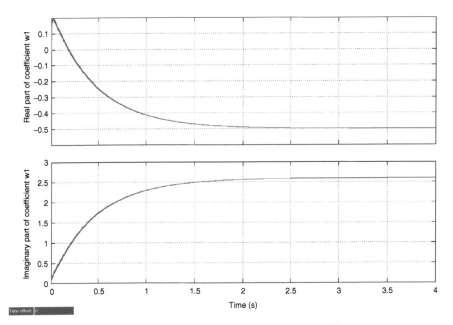

Figure 13.36 Real and Imaginary Parts of the Coefficient w_1.

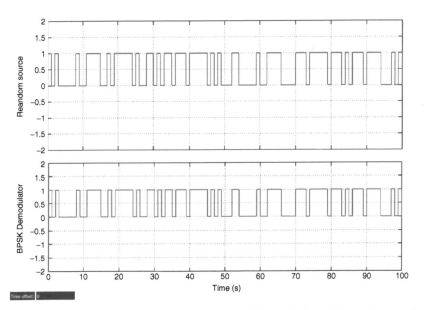

Figure 13.37 Random Source Output (Top) and Demodulator Output (Bottom).

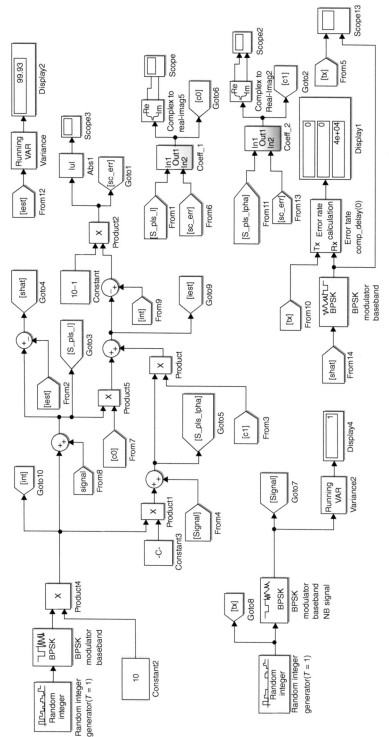

Figure 13.38 Noise Free Simulink Model of Interference Nulling, $S/I = -10\,\text{dB}$.

The Simulink model parameters for this example are now listed as follows:

Model Parameters for Antenna Nulling System with Interference and LMS Cancellation

- Sample-based simulation
- Simulation time = 40,000 s
- Data symbol sample time = 1 s
- Interference symbol sample time = 1 s
- Interference incident angle $\phi_i = 45°$
- Interference constant $C = \exp(-j3\pi\cos(\pi/4))$
- Interference amplitude = 10V, $S/I = -10$ dB
- BPSK modulation for data and interference
- Data source seed = 37
- Interference source seed = 23
- No additive noise
- Two tap LMS algorithm with scale factor = 0.1

The error rate calculation shown in Figure 13.38 indicates that no errors were made with $S/I = -10$ dB over the 40,000 s simulation period. Figure 13.39 shows a comparison of the source and demodulator outputs from scope13 again showing effective interference nulling.

13.5 KALMAN FILTERING

The Kalman filter is a recursive algorithm known to be the optimum minimum MSE estimator in Gaussian noise. The Kalman filter can be viewed as a generalization of the recursive lease squares algorithm for stochastic signals and applies when the signals and noise are nonstationary. A brief summary of the Kalman algorithm is presented here, with vectors and matrices displayed in bold font.

The L-dimensional state vector of a system $\vec{u}_k$ at time sample k is expressed in terms of its prior state at time sample $k-1$ according to

$$\vec{u}_k = A(k, k-1)\vec{u}_{k-1} + B_{k-1}\vec{\xi}_{k-1}$$

where $A(k, k-1)$ is a known $L \times L$ state transition matrix, B_{k-1} is a known $L \times P$ model noise matrix, and ξ_{k-1} is a zero mean noise vector often referred

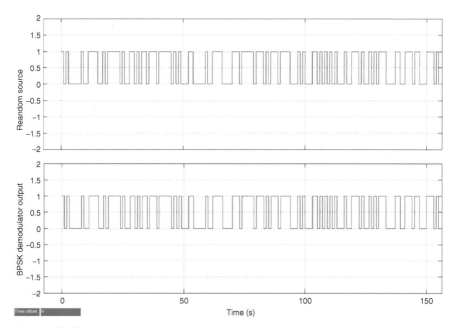

Figure 13.39 Random Source Output (Top) and Demodulator Output (Bottom) with $S/I = -10\,\text{dB}$.

to as "plant" noise. Here $\vec{\xi}_k$ is assumed to be normally distributed, denoted by $N(0, \ \Xi_k)$ where $\Xi_k = E[\vec{\xi}_k \vec{\xi}_k^*]$ is the covariance of $\vec{\xi}_k$. The measurement model is expressed by an N-dimensional vector $\vec{y}_k$ given by

$$\vec{y}_k = H_k \vec{u}_k + \vec{n}_k$$

where H_k is an $N \times L$ channel transition matrix and $\vec{z}_k$ is an N-dimensional zero mean channel noise vector. Here $\vec{z}_k$ is independent of $\vec{\xi}_k$ and is normally distributed $N(0, \ V_k)$ with covariance $V_k = E[\vec{n}_k \vec{n}_k^*]$.

The current estimate of $\vec{u}_k$ is denoted by $\hat{\vec{u}}_k$ and given by

$$\hat{\vec{u}}_k = A(k, \ k-1)\hat{\vec{u}}_{k-1} + G_k \ (\vec{y}_k - H_k A(k, \ k-1)\hat{\vec{u}}_{k-1})$$

where G_k is referred to as the Kalman gain. The error associated with the current signal, $\vec{e}_k$, is given by

$$\vec{e}_k = \vec{u}_k - \hat{\vec{u}}_k$$

Denoting $\vec{\hat{u}}_{k,k-1} = A(k, k-1)\vec{\hat{u}}_{k-1}$ as the prediction of the current signal, the current error covariance matrix C_k is defined as

$$C_k = E\{(\vec{u}_k - \vec{\hat{u}}_{k,k-1})(\vec{u}_k - \vec{\hat{u}}_{k,k-1})^{T*}\}$$

The current error covariance matrix is determined recursively from the predicted error covariance matrix $C_{k,k-1}$ according to

$$C_k = C_{k,k-1} - G_k H_k C_{k,k-1}$$

where

$$C_{k,k-1} = A(k, k-1) C_{k-1} A(k, k-1)^{T*} + B_{k-1} \Xi_{k-1} B_{k-1}^{T*}$$

The Kalman gain is obtained from

$$G_k = C_{k,k-1} H_k^{T*} [H_k C_{k,k-1} H_k^{T*} + V_k]^{-1}$$

A summary of the principal expressions is provided in Table 13.3.

13.5.1 Scalar Kalman Filter

A simple, scalar version of the Kalman filter is presented next. The system dynamic model is given by

$$u_k = a u_{k-1} + \xi_{k-1}$$

TABLE 13.3 Summary of the Kalman Algorithm

System dynamic model	$\vec{u}_k = A(k, k-1)\vec{u}_{k-1} + B_{k-1}\vec{\xi}_{k-1}$
Measurement model	$\vec{y}_k = H_k \vec{u}_k + \vec{n}_k$
Current signal estimate	$\vec{\hat{u}}_k = A(k, k-1)\vec{\hat{u}}_{k-1}$
	$\quad + G_k (\vec{y}_k - H_k A(k, k-1)\vec{\hat{u}}_{k-1})$
Error of current signal	$\vec{e}_k = \vec{u}_k - \vec{\hat{u}}_k$
Prediction of current signal	$\vec{\hat{u}}_{k,k-1} = A(k, k-1)\vec{\hat{u}}_{k-1}$
Current Error covariance matrix	$C_k = C_{k,k-1} - G_k H_k C_{k,k-1}$
Predicted error covariance matrix	$C_{k,k-1} = A(k, k-1) C_{k-1} A(k, k-1)^{T*}$
	$\quad + B_{k-1} \Xi_{k-1} B_{k-1}^{T*}$
Kalman gain	$G_k = C_{k,k-1} H_k^{T*} [H_k C_{k,k-1} H_k^{T*} + V_k]^{-1}$

KALMAN FILTERING

where $A(k, k-1) = a$ and $B_{k-1} = 1$. The measurement model is reduced to

$$y_k = hu_k + n_k$$

where $H_k = h$. The noise terms are independent normal variables where the distribution for ξ_{k-1} is $N(0, \sigma_\xi^2)$ and the distribution for n_k is $N(0, \sigma_n^2)$. The current signal estimate is

$$\hat{u}_k = a\hat{u}_{k-1} + G_k(y_k - ha\hat{u}_{k-1})$$

The predicted error covariance is

$$C_{k,k-1} = a^2 C_{k-1} + \sigma_\xi^2$$

and the current error covariance is

$$C_k = (a^2 C_{k-1} + \sigma_\xi^2)(1 - G_k h)$$

The Kalman gain is then expressed as

$$G_k = \frac{(a^2 C_{k-1} + \sigma_\xi^2)h}{h^2 a^2 C_{k-1} + h^2 \sigma_\xi^2 + \sigma_n^2}$$

A numerical calculation is now performed using $a = 0.5$, $h = 1$, $\sigma_\xi^2 = 2$, $\sigma_n^2 = 1$, and with the initial conditions $C_{-1} = 1$ and $\hat{u}_{-1} = 0$. In this case $G_k = C_k$. Substituting the numerical values for $k = 0$ results in

$$\hat{u}_{0,-1} = \frac{1}{2}\hat{u}_{-1} = 0$$

$$C_{0,-1} = \left(\frac{1}{2}\right)^2 C_{-1} + 2 = \frac{9}{4}$$

$$G_0 = \frac{\left[\left(\frac{1}{2}\right)^2(1) + 2\right](1)}{1^2\left(\frac{1}{2}\right)^2(1) + 1^2(2) + 1} = \frac{9}{13} = C_0$$

$$\hat{u}_0 = \frac{1}{2}\hat{u}_{-1} + \frac{9}{13}\left(y_0 - \frac{1}{2}\hat{u}_{-1}\right) = \frac{9}{13}y_0$$

For $k = 1$ the results are

$$\hat{u}_{1,0} = \frac{1}{2}\hat{u}_0$$

$$C_{1,0} = \left(\frac{1}{2}\right)^2 C_0 + 2 = \frac{113}{52}$$

$$G_1 = \frac{\left[\left(\frac{1}{2}\right)^2 \left(\frac{9}{13}\right) + 2\right](1)}{1^2\left(\frac{1}{2}\right)^2 \left(\frac{9}{13}\right) + 1^2(2) + 1} = \frac{113}{165} = C_1$$

$$\hat{u}_1 = \frac{1}{2}\hat{u}_0 + \frac{113}{165}\left(y_1 - \frac{1}{2}\hat{u}_0\right) = \frac{113}{165}y_1 + \frac{6}{55}y_0$$

The Simulink model for the scalar Kalman example is presented in Figure 13.40 and the parameters for this example are listed as follows:

Model Parameters for Scalar Kalman Filter

- Sample-based simulation with 1s sample time
- Simulation time = 100 s
- System dynamic model $u_k = au_{k-1} + \xi_{k-1}, a = 0.5, \xi_{k-1}$ is $N(0, \sigma_\xi^2)$, $\sigma_\xi^2 = 2$
- Measurement model $y_k = hu_k + n_k, h = 1, n_k$ is $N(0, \sigma_n^2), \sigma_n^2 = 1$
- Initial conditions $C_{-1} = 1, \hat{u}_{-1} = 0$
- Current estimate $\hat{u}_k = a\hat{u}_{k-1} + G_k(y_k - ha\hat{u}_{k-1})$
- Predicted error covariance: $C_{k,k-1} = a^2 C_{k-1} + \sigma_\xi^2$
- Current error covariance: $C_k = (a^2 C_{k-1} + \sigma_\xi^2)(1 - G_k h)$
- Kalman gain $G_k = \frac{(a^2 C_{k-1} + \sigma_\xi^2)h}{h^2 a^2 C_{k-1} + h^2 \sigma_\xi^2 + \sigma_n^2}$

The Simulink model displayed in Figure 13.40 includes an explicit representation of the computations based on the expressions presented here along with a Kalman filter Simulink library routine enabling a direct comparison. The Kalman gain, G_k, the current error covariance, C_k, the measurement model output, y_k, and the current estimate, $\hat{u}_k$, data are sent to the workspace where a computation will allow verification. Table 13.4 displays the workspace results for $k = 0$ and 1.

Substituting $y_0 = -0.6492$ into $\hat{u}_0 = \frac{9}{13}y_0$ results in $\hat{u}_0 = -0.4494$; similarly substituting the results for y_0 and y_1 into $\hat{u}_1 = \frac{113}{165}y_1 + \frac{6}{55}y_0$ results in $\hat{u}_1 = 1.3280$. In other words, the Simulink output in the last column in Table 13.4 agrees with the numerical computations.

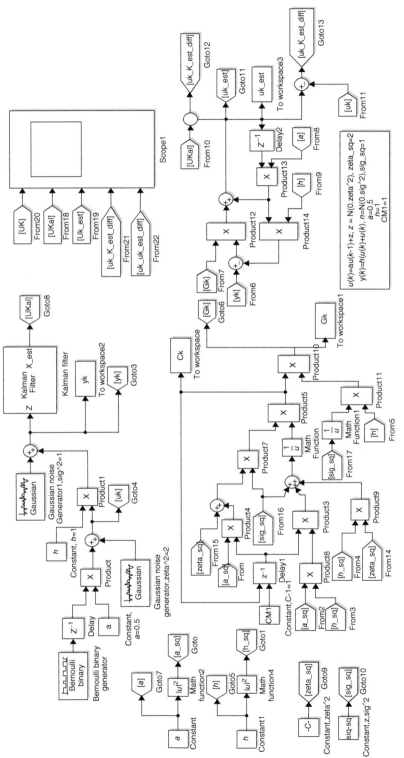

Figure 13.40 Scalar Kalman Simulink Model.

TABLE 13.4 Workspace Values for Scalar Kalman Example

k	G_k	C_k	y_k	$\hat{u}_k$
0	0.6923	0.6923	−0.6492	−0.4494
1	0.6848	0.6848	2.0426	1.3280

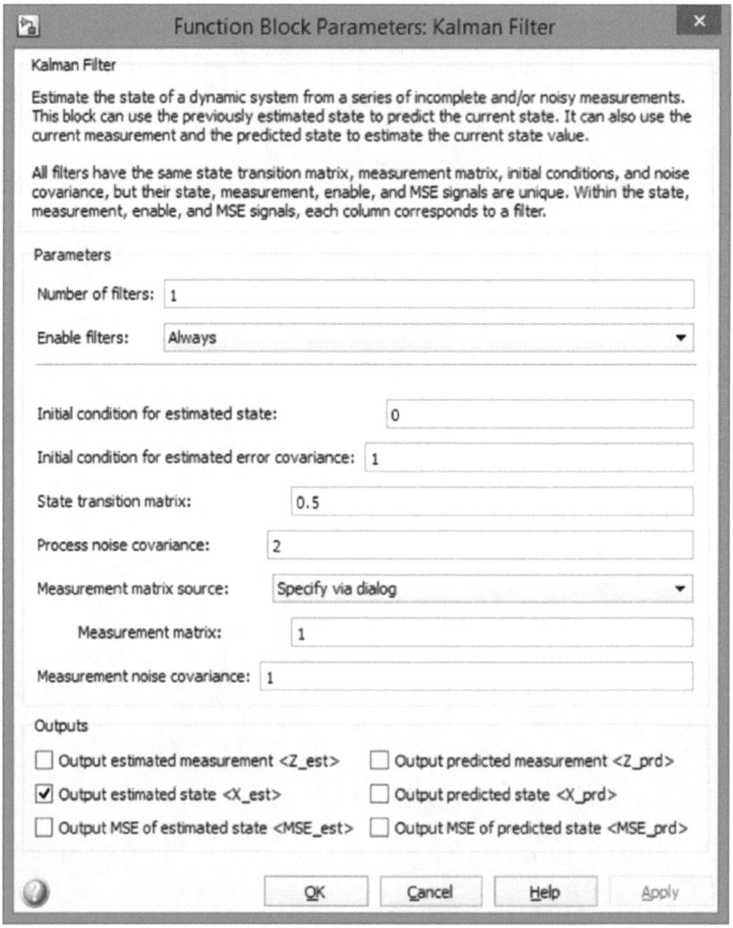

Figure 13.41 Scalar Kalman Input Parameters.

Now the results of using the Kalman Simulink library block will be compared with the explicit representation in the Simulink model. Figure 13.41 displays the window for entering the Kalman parameters that are seen to be the same as those used in the direct Simulink computations.

KALMAN FILTERING

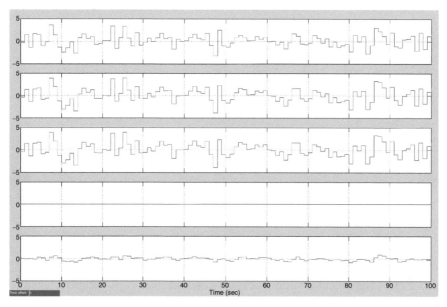

Figure 13.42 Scope Display for Scalar Kalman Example.

Figure 13.42 displays the scope output for the various signals from top to bottom as follows:

- Top trace: u_k, system dynamic model output
- Second down from top: Kalman filter Simulink library output of current estimate, $\hat{u}_k$,
- Third trace down from top: direct Kalman computation for current estimate, $\hat{u}_k$,
- Fourth trace down from top: difference between the Simulink library output and the direct computation of the current estimate, $\hat{u}_k$
- Bottom trace: difference between the direct computation of the current estimate, $\hat{u}_k$ and the system dynamic model u_k.

By examining the fourth trace down shown in Figure 13.42, it is observed that the results of the current estimate from the Kalman Simulink library and the results from the explicit representation in the Simulink model are in agreement. The bottom trace shown in Figure 13.42 indicates that the Kalman algorithm exhibits a residual error.

13.5.2 Kalman Equalizer

In this example, the Kalman filter is used to compute the equalizer coefficients required to mitigate the distortion introduced by a multipath channel. For simplicity, all variables are treated as real and no additive noise is present. The Kalman filter equalizer coefficients, identified as an L-dimensional vector $\vec{d}_k$, are computed recursively as follows

$$\vec{d}_k = \vec{d}_{k-1} + G_k(s_k - \vec{v}_k^T \vec{d}_{k-1})$$

where s_k is a known training signal, $\vec{v}_k$ is the multipath corrupted signal at the channel output, and G_k is the Kalman gain. The current error covariance matrix is computed from

$$C_k = C_{k-1} + Q_k - G_k \vec{v}_k^T (C_{k-1} + Q_k)$$

where Q_k is the dynamic model noise covariance and V_{ke} is the covariance of the channel model noise. The Kalman gain is then given by

$$G_k = (C_{k-1} + Q_k)\vec{v}_k / [\vec{v}_k^T (C_{k-1} + Q_k)\vec{v}_k + V_{ke}]$$

where $Q_k = 0.01\mathbf{I}$ is the assumed initial value and $\mathbf{I}$ denotes the $L \times L$ identity matrix. The initial value for $V_{ke} = 0.1$.

The simulation presented next assumes BPSK modulation, a two-path multipath channel, a tapped delay line equalizer with five tap coefficients computed using the Kalman filter equations that were just given.

In the simulations that follow a MATLAB function for the Kalman coefficient computation is incorporated into the model, rather than using an explicit representation of the computations in Simulink. This example demonstrates the utility of embedding a MATLAB routine within Simulink when a Simulink library function is unavailable. In this case, the Simulink model for the Kalman equalizer example, displayed in Figure 13.43, has the parameters listed as follows:

Model Parameters for Kalman Equalizer

- Sample-based simulation with 1 s sample time
- BPSK modulation
- Simulation time = 1000 s

- Data symbol sample time = 1 s
- Multipath delay = 1 s
- Multipath gain = 0.5
- No additive noise
- Signals are real
- 5 equalizer taps spaced at 1s intervals
- Kalman equalizer coefficients computed in MATLAB function, kalman_equal
- Initial conditions: $Q_k = 0.01\mathbf{I}$, $V_{ke} = 0.1$.

The MATLAB function routine is provided in Figure 13.44.

In the MATLAB function, the MSE E_N at step N, is computed recursively according to

$$E_N = \frac{1}{N}\sum_{m=1}^{N} e_m^2 = \frac{N-1}{N}E_{N-1} + \frac{1}{N}e_N^2$$

where e_m is the error at step m.

In the Figure 13.43 Simulink model, it can be seen that the training signal is delayed by 2 s in order to synchronize the equalizer with the center tap. For the parameter selections used, Figure 13.43a indicates that the BER ≈ 0.002 and confirms that the equalizer is effective in mitigating the multipath.

Figure 13.45 displays a part of the simulation for the BPSK modulator input and the BPSK demodulator output where the two signals are mostly in agreement due to the low BER.

Figure 13.46 shows the real part of the BPSK modulator output along with the Kalman equalizer output. Between 0 and 10 s, the equalizer is converging and errors are to be expected in that interval.

The signals for the Kalman equalizer coefficients are displayed in Figure 13.47. The center tap is seen to have the largest magnitude.

Thus far in this chapter, the benefits of using a Kalman filter equalizer are not apparent, particularly since an LMS equalizer has a simpler implementation. It is therefore useful to compare the Kalman and LMS equalizer performance.

Figure 13.48 displays a Simulink model for an LMS equalizer that mitigates the distortion introduced by the two-path channel model shown in Figure 13.43b. A MATLAB function is used to compute the LMS equalizer coefficients.

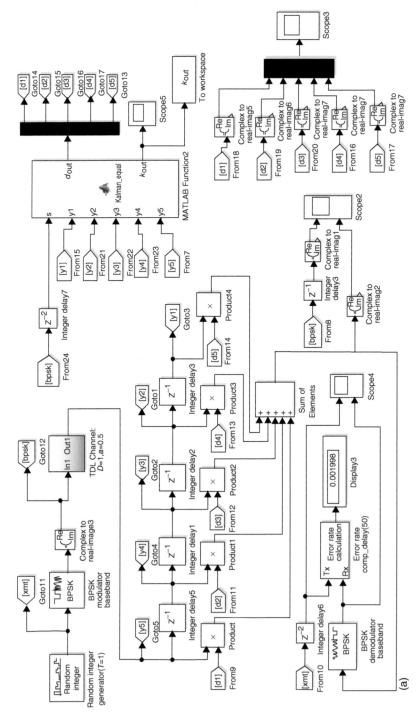

Figure 13.43 (a) Simulink Model for Kalman Equalizer; (b) Multipath Channel Used with the Kalman Equalizer.

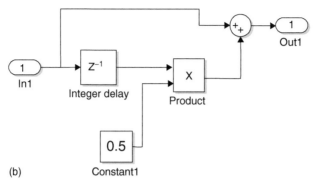

(b)

Figure 13.43 *(Continued)*

The Simulink model parameters for the LMS equalizer example, displayed in Figure 13.48, are listed as follows:

Model Parameters for LMS Equalizer

- Sample-based simulation with 1 s sample time
- BPSK Modulation
- Simulation time = 1000 s
- Data symbol sample time = 1 s
- Multipath delay = 1 s
- Multipath gain = 0.5
- No additive noise
- Signals are real
- 5 equalizer taps spaced at 1s intervals
- LMS equalizer coefficients computed in MATLAB function, lms_eq
- LMS scale factor del = 0.01

The MATLAB function routine for the LMS equalizer coefficients is provided in Figure 13.49.

In the Figure 13.48 Simulink model, the training signal is delayed by 2 s in order to synchronize the equalizer at the center tap. For the selected parameters, the BER is about 0.02, which indicates that the LMS equalizer mitigates the multipath but not as well as the Kalman equalizer.

Figure 13.50 displays a part of the simulation for the BPSK modulator input and the BPSK demodulator output. The slower rate of LMS equalizer convergence is observed where agreement between the two signals is achieved after about 100 s.

Figure 13.51 displays the real part of the BPSK modulator output along with the Kalman equalizer output. The slow rate of LMS equalizer convergence introduces errors as expected.

MATLAB Function for Kalman Equalizer Coefficients

```
%Kalman Equalizer Coefficient Computation
%Inputs
    %Outputs of multipath channel in delay line, y1...y5
    %Training symbols,s
%Outputs:
    %Kalman equalizer tap coefficient,d
    %Squared magnitude of Kalman error,kout
%Comments:
    %Kalman equalizer coefficients computed
    %Kalman gain,g
    %Current error covariance,cov
    %Initialization
        % equalizer coefficients,dout
        % channel noise covariance,vke
        % dynamic model(plant) noise covariance,q
        % initial current error covariance,cov
    %Delay to fill equalizer,nn
function [dout, kout] =kalman_equal(s,y1,y2,y3,y4,y5)
persistent d
persistent nn
persistent cov
persistent err_sq
% Initialization
if isempty(d)
    d =[.5+0.*i .5+0.*i .5+0.*i .5+0.*i .5+0.*i]';
    cov=0*eye(5);
    nn=0;
    err_sq=0;
end;
if nn>3
    vke=.1;
    q=.01*eye(5);
    v=[y5 y4 y3 y2 y1]';
    e=s-d'*v;
    err_sq=err_sq*((nn-1)/nn)+(abs(e))^2/nn;
    cq=cov+q;
    den=v'*cq*v+vke;
    g=cq*v/den;
    term=g*v'*cq;
    cov=real(cq-g*v'*cq);
    d=d+g*e;
else
    nn=nn+1;
    cov=0*eye(5);
    err_sq=0;
end;
nn=nn+1;
kout=err_sq;
dout=d;
```

Figure 13.44 Kalman Equalizer Coefficient Computation.

KALMAN FILTERING

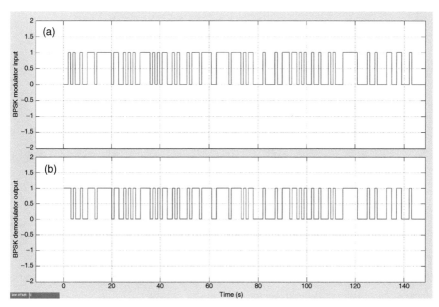

Figure 13.45 BPSK Input (a) and BPSK Demodulator Output (b) for Kalman Equalizer.

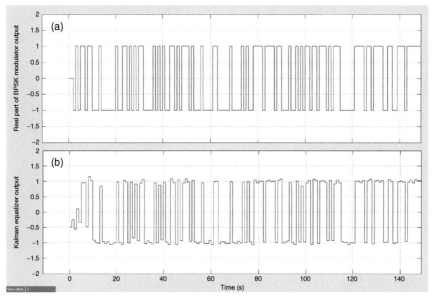

Figure 13.46 Real Part of BPSK Modulator Output (a) and Kalman Equalizer Output (b).

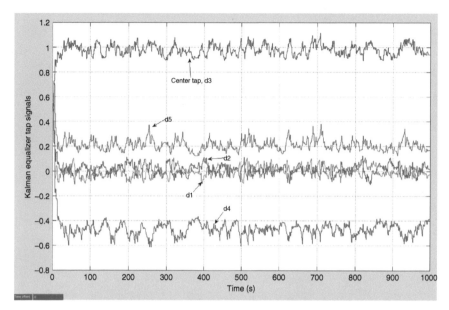

Figure 13.47 Kalman Equalizer coefficients.

The signals for the LMS equalizer coefficients are displayed in Figure 13.52, where it is observed that the center tap has the largest magnitude.

In the case of an LMS equalizer, the equalizer coefficients can be computed using the equations developed in Chapter 12. For a real multipath component a, the coefficients are obtained as follows

$$\begin{bmatrix} (1+a^2) & a & 0 & 0 & 0 \\ a & (1+a^2) & a & 0 & 0 \\ 0 & a & (1+a^2) & a & 0 \\ 0 & 0 & a & (1+a^2) & a \\ 0 & 0 & 0 & a & (1+a^2) \end{bmatrix} \begin{bmatrix} c_5 \\ c_4 \\ c_3 \\ c_2 \\ c_1 \end{bmatrix} = \begin{bmatrix} 0 \\ 0 \\ 1 \\ a \\ 0 \end{bmatrix}$$

Substituting $a = 0.5$ in the aforementioned equation and solving for the coefficients yields

$$\begin{bmatrix} c_5 \\ c_4 \\ c_3 \\ c_2 \\ c_1 \end{bmatrix} = \begin{bmatrix} 0.1875 \\ -0.4689 \\ 0.9846 \\ 0.0073 \\ -0.0029 \end{bmatrix}$$

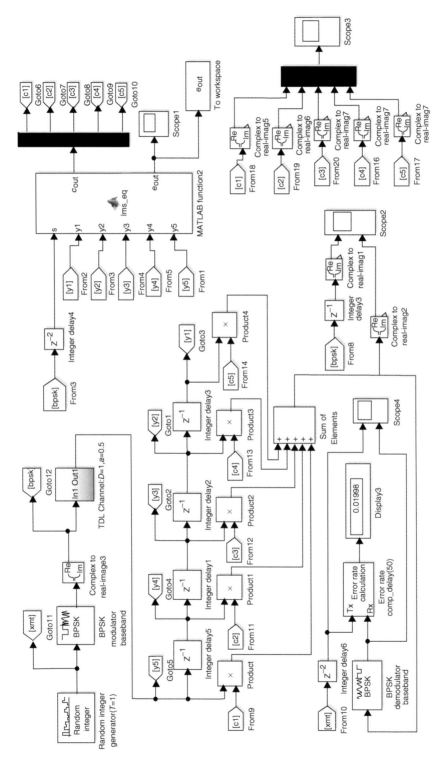

Figure 13.48 Simulink Model for LMS Equalizer.

MATLAB Function for LMS Equalizer Coefficients

```
%LMS Equalizer Coefficient Computation
%Inputs
   %Outputs of multipath channel in delay line, y1...y5
   %Training symbols, s
%Outputs:
   %Equalizer tap coefficient,cout
   %Squared magnitude of LMS error,eout
%Comments:
   %LMS equalizer coefficients computed
   %Initialization of equalizer coefficients
   %LMS scale factor del=0.01
   %Delay to fill equalizer,nn
% Initialization
function [cout, eout] = lms_eq(s, y1, y2, y3, y4, y5)
persistent c
persistent nn
persistent err_sq
if isempty(c)
    c =[.5+0.*i .5+0.*i .5+0.*i .5+0.*i .5+0.*i]';
    nn=0;
    err_sq=0;
end;
if nn>3
    del=.01;
    v=[y5 y4 y3 y2 y1]';
    esd=s-c'*v;
    err_sq=err_sq*((nn-1)/nn)+(abs(esd))^2/nn;
    c=c+del*esd*v;
else
    nn=nn+1;
    err_sq=err_sq;
end;
nn=nn+1;
eout=err_sq;
cout=c;
```

Figure 13.49 LMS Equalizer Coefficient Computation.

It can be seen that the numerical values obtained above are being approached in both Figure 13.47 for the Kalman equalizer coefficients and in Figure 13.52 for the LMS equalizer coefficients.

Another comparison between the Kalman and LMS equalizer performance can be made by examining the MSE results shown in Figure 13.53, which reveals that the Kalman equalizer converges more rapidly than the LMS equalizer.

KALMAN FILTERING

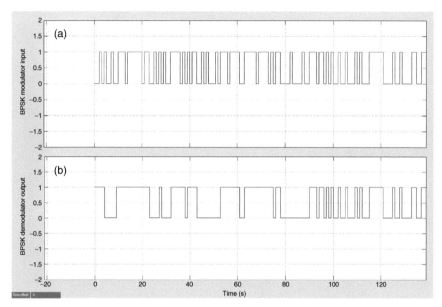

Figure 13.50 BPSK Input (a) and BPSK Demodulator Output (b) for LMS Equalizer.

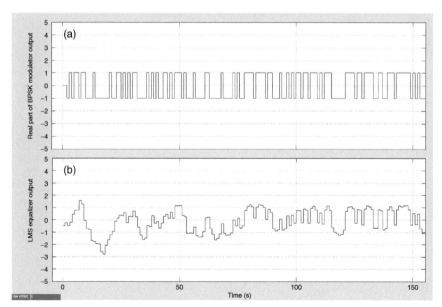

Figure 13.51 Real Part of BPSK Modulator Output (a) and LMS Equalizer Output (b).

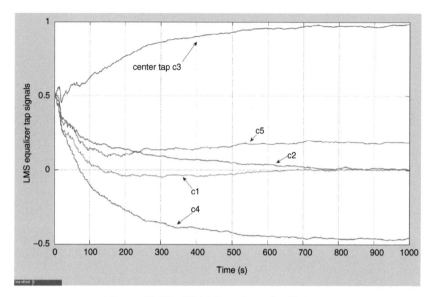

Figure 13.52 LMS Equalizer Coefficients.

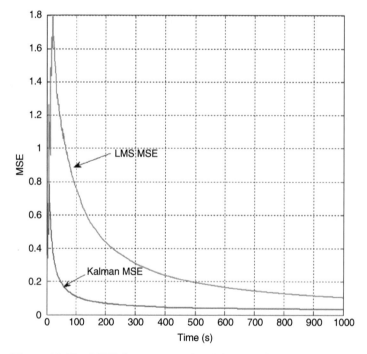

Figure 13.53 MSE Convergence for Kalman and LMS Equalizers.

13.5.3 Radar Tracking Using Extended Kalman Filter (EKF)

A more traditional example of Kalman filtering is now presented where an aircraft is tracked using an EKF.[2] The EKF equations include a linearized system dynamic model, represented by a 4×4 matrix A with a covariance q, a linearized measurement matrix h with a measurement model covariance vke and an estimate of the location xhat. The Simulink model for the radar tracking is shown in Figure 13.54.

In the Simulink model, the aircraft motion is obtained by integrating the acceleration of a noisy input followed by successive integrations to obtain aircraft velocity and position. The EKF input consists of a noisy measurement added to the aircraft motion. The EKF is computed by means of a MATLAB function identified as EKF.

The MATLAB model parameters are listed as follows:

Model Parameters for Aircraft Tracking Example

- Sample-based simulation with 0.1 s sample time
- Simulation time = 150 s
- Aircraft motion parameters
  - Acceleration with two parameters: 1/5, 1/4
  - Velocity: 182.3 ft/s = 200 km/h
  - Noise power; [0.5 0.5]
- Measurement parameters
  - Noise power: [1 1]
  - Measurement gain: [500 0; 0 500], [5000 0; 0 5000]
- Model matrix A: [1 1 0 0; 0 1 0 0; 0 0 1 1; 0 0 0 1]
- Model covariance, q: diag([0 0.005 0 0.005])
- Measurement model covariance, vke: diag([8^2 1^2])
- North-south aircraft location: xhat(1)
- East-west aircraft location: xhat(3)
- MATLAB function: EKF.

Figure 13.55 displays the MATLAB routine EKF.

Figure 13.56 displays the results of aircraft tracking using an EKF. In this figure, the aircraft flies at 200 km/h and is subjected to a measurement noise gain matrix [500 0; 0 500].

---

[2] EKF equations are presented in Schonhoff and Giordano, op. cit., Chapter 14

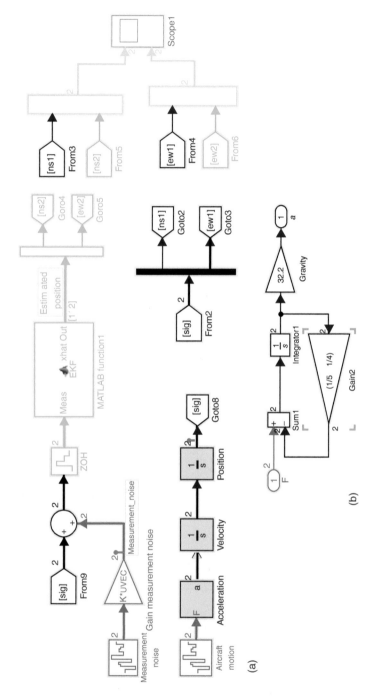

Figure 13.54 (a) Simulink Model for Radar Tracking of an Aircraft; (b) Aircraft Acceleration Model.

Figure 13.57 displays the same result as that in Figure 13.56, now with the noise measurement gain matrix increased by a factor of 10, to [5000 0; 0 5000].

Figures 13.56 and 13.57 results for aircraft tracking show that the EKF is effective in determining the aircraft location even in the presence of increased measurement noise.

MATLAB Function for EKF

```
function xhatOut  = EKF(meas)
%Extended Kalmann Filter to Track Aircraft Using Radar
%Inputs
   % Measurement noise with noise covariance,vke
   % Aircraft motion with dynamic model covariance q
   % Aircraft speed =182.3ft/sec=200km/hr
%Outputs:
    %Estimated Aircraft Location
    %xhat(1)=north south
    %xhat(3)=east west
%Comments:
    %Adapted from MathWorks demo sldemo_radar_eml
% Initialization
persistent cov;
persistent xhat
if isempty(cov)
    xhat = [0.001; 0.01; 0.001; 182.3];
    cov = zeros(4);
end
dt=1;

% Model parameters A, q, and vke
A = [1 dt 0 0; 0 1 0 0 ; 0 0 1 dt; 0 0 0 1];
q =   diag([0 .005 0 .005]);
vke =   diag([ 8^2 1^2]);
% Predict the covariance matrix:
cov = A*cov*A' + q;
% Predict the track estimate::
xhat = A*xhat;
range = sqrt(xhat(1)^2+xhat(3)^2);
%  bearing = atan2(xhat(3),xhat(1));
% Compute linearized measurement matrix h
h = [ xhat(1)/range         0 xhat(3)/range          0
   -xhat(3)/range     0  xhat(1)/range      0 ];
% Compute estimation error
err = meas - [xhat(1) xhat(3)].';
% Compute Kalman gain:
g = cov*h'*inv(h*cov*h'+ vke);
%  Current estimate
xhat = xhat + g*err;
% Current covariance matrix
cov = (eye(4)-g*h)*cov*(eye(4)-g*h)' + g*vke*g';
```

Figure 13.55 MATLAB Function EKF for Aircraft Tracking.

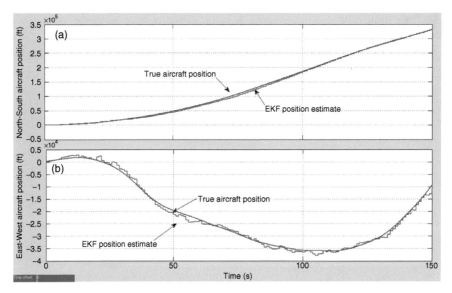

Figure 13.56 Aircraft Tracking Using EKF: North–South (a), East–West (b).

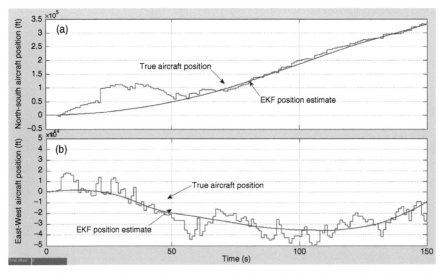

Figure 13.57 Aircraft Tracking Using EKF: North–South (a), East–West (b); Measurement Noise Gain Increased by a Factor of 10.

13.6 ORTHOGONAL FREQUENCY DIVISION MULTIPLEXING

Orthogonal frequency division multiplexing (OFDM) is a spectrally efficient modulation technique employed in many military and commercial communications systems. Some of the commercial applications include IEEE802.11 variants (a, g, n, etc.) adopted for wireless local area networks, IEEE802.16 (Worldwide Interoperability for Microwave Access or WIMAX), Long Term Evolution (LTE) for mobile phone communications, and digital audio broadcasting. By applying data from QAM modulation onto orthogonal, closely spaced carriers, each with low symbol rate, and appending guard time, high overall data rate is attained while mitigating intersymbol interference. A cyclic prefix inserted during the guard interval assists in minimizing the time dispersion inherent in multipath channels. OFDM is efficiently implemented using the fast Fourier transform (FFT), which enables the use of frequency domain equalization, an additional scheme for mitigating multipath. OFDM is sensitive to Doppler shift and exhibits high peak to average power ratio requiring transmission to be maintained within the linear range of the power amplifier.

The Simulink model to be presented here is intended to explore selected characteristics of OFDM. The model, depicted in Figure 13.58 for an AWGN channel, is modified from an IEEE 802.11a Simulink model available on the MATLAB Central File Exchange.[3]

Since it is known that OFDM BER performance often corresponds to 16-QAM, a simulation for 16-QAM in AWGN is included in the model for comparison with the OFDM performance. The scope is used to ensure proper synchronization between the transmitted and received signals.

Each of the blocks in this model is now discussed further to explain several essential elements of an OFDM implementation. The source block produces frame-based, 16-ary random integers with 960 samples/frame as seen in Figure 13.59.

Selection of the 16-ary rectangular QAM modulator with Gray mapping and average power normalization is shown in Figure 13.60a; the signal constellation is displayed in Figure 13.60b.

Figure 13.61 displays the shaping of OFDM symbols, found by looking under the mask, where it is seen that the 960 samples/frame are converted into a 48×20 array.

To create an efficient FFT size, the Pad block, shown in Figure 13.62, appends zeroes to increase the column size to 64.

---

[3] http:// www.mathworks.com/matlabcentral

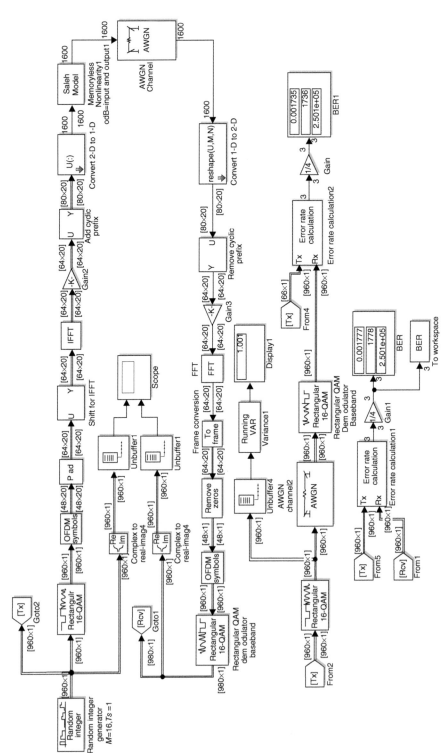

Figure 13.58 OFDM Simulink Model ($E_b/N_o = 10$ dB).

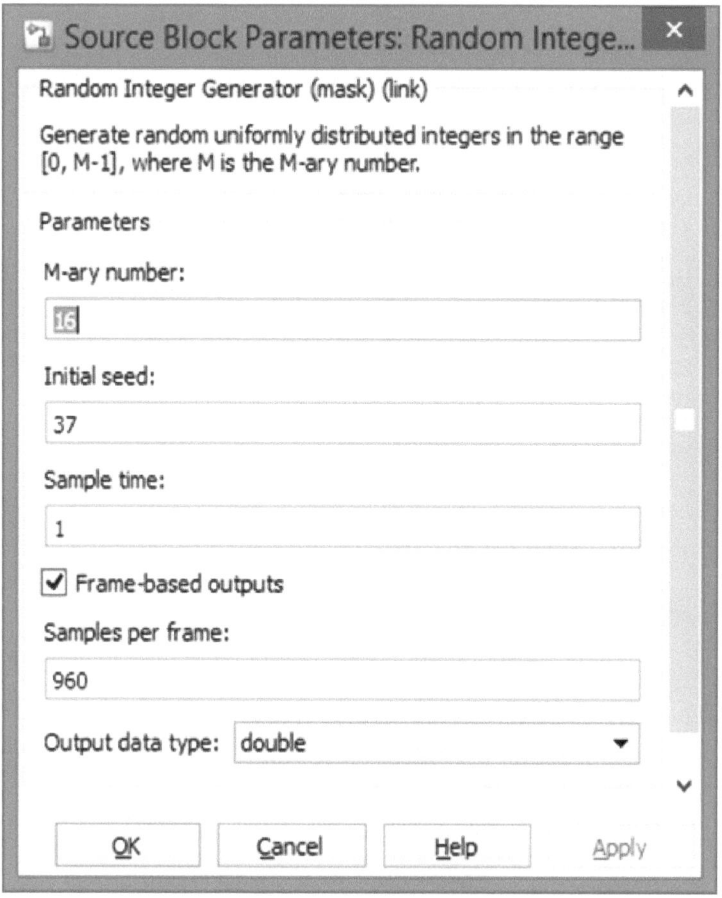

ORTHOGONAL FREQUENCY DIVISION MULTIPLEXING 345

Figure 13.59 Source Block Generation of 16-ary Random Integers.

The selector block, labeled "shift for IFFT" as shown in Figure 13.63, modifies the indices of the array for input to the Inverse Fast Fourier transform (IFFT).

With the column size having been set at 64 in Figure 13.62, the OFDM modulation is implemented using a 64-point IFFT block depicted in Figure 13.64.

The gain block applies the normalization as $1/\sqrt{64} = 1/8$.

The "Add Cyclic Prefix" block, whose parameters are shown in Figure 13.65, appends 16 symbols for the cyclic prefix and expands the output array to 80×20.

The block labeled "Convert 2-D to 1-D" modifies the 80×20 array to a single dimension for input to the Saleh model, which characterizes the

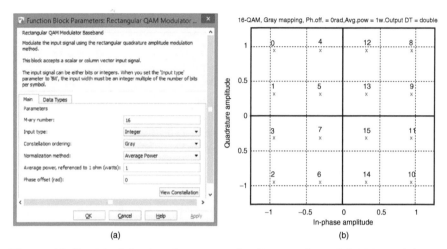

Figure 13.60 (a) Selecting Parameters for Rectangular 16-QAM Modulator; (b) Rectangular 16-QAM Constellation.

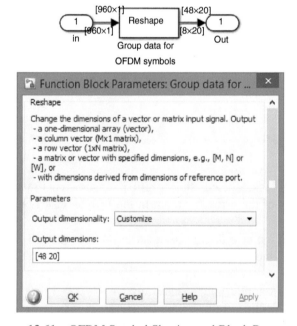

Figure 13.61 OFDM Symbol Shaping and Block Parameters.

Figure 13.62 Zero Pad Block Parameters.

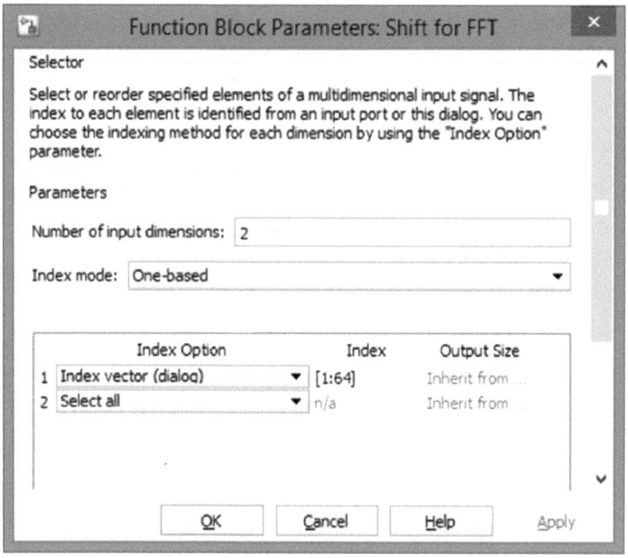

Figure 13.63 Selector Block, Shift for IFFT.

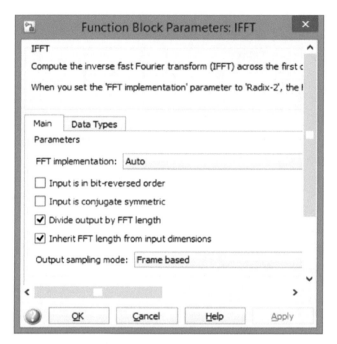

Figure 13.64 IFFT for OFDM Implementation.

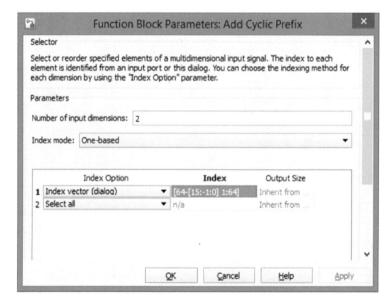

Figure 13.65 "Add Cyclic Prefix" Block Parameters.

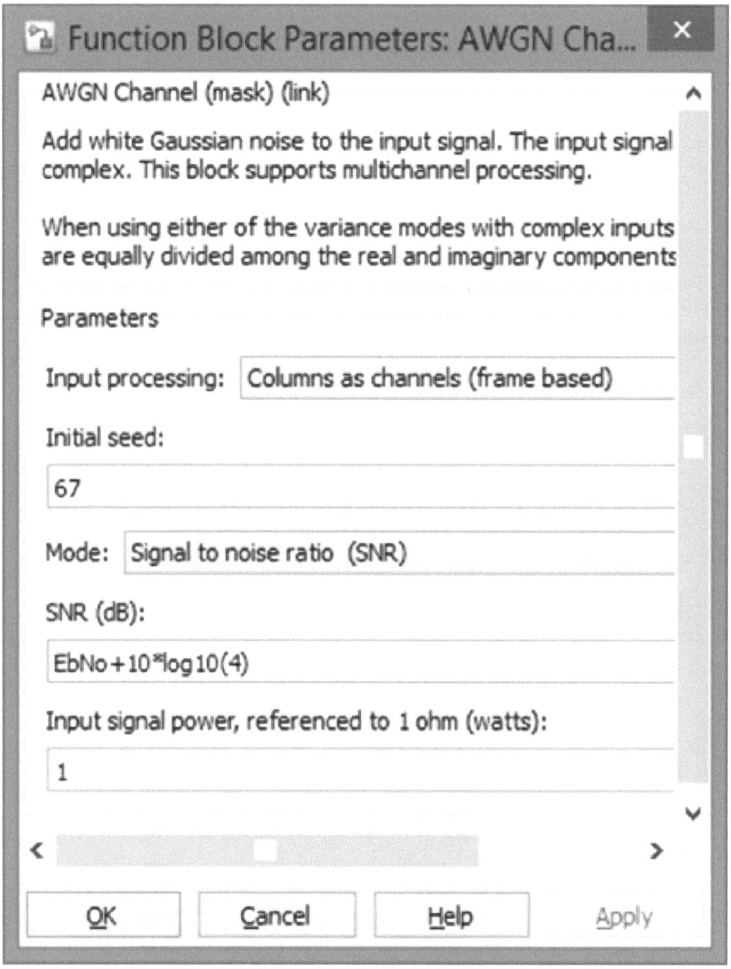

ORTHOGONAL FREQUENCY DIVISION MULTIPLEXING

Figure 13.66 AWGN Parameters for OFDM Simulation.

power amplifier. The Saleh model is operated with AM/AM = [1 0] and AM/PM = [0 0], thereby forcing the OFDM signal to remain within the linear region of the power amplifier. The setting of the AWGN block parameters for this simulation is shown in Figure 13.66, where $E_b/N_o = 10$ is specified in the MATLAB command window.

The receive path converts the single dimensional vector back to an 80×20 array for insertion into the "Remove Cyclic Prefix" block, whose parameters are displayed in Figure 13.67.

Following the gain normalization as $1/\sqrt{64} = 1/8$, a 64-point FFT is performed with parameters shown in Figure 13.68.

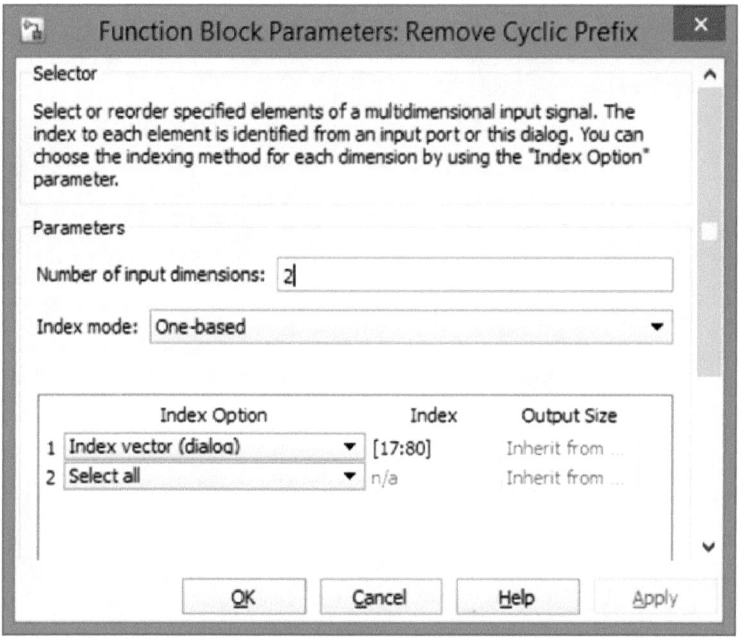

Figure 13.67 "Remove Cyclic Prefix" Block Parameters.

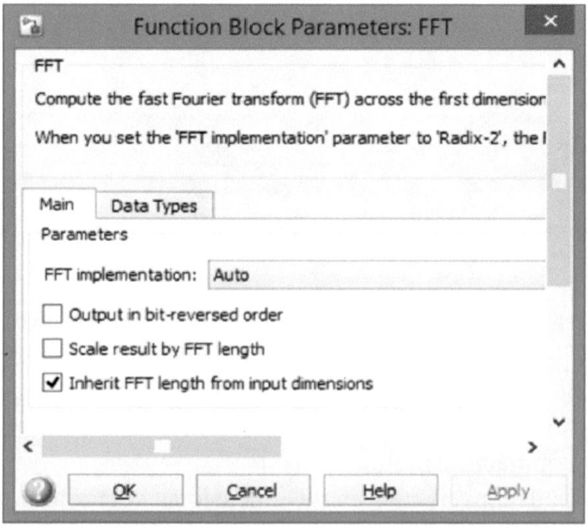

Figure 13.68 Parameters for 64-Point FFT.

ORTHOGONAL FREQUENCY DIVISION MULTIPLEXING 351

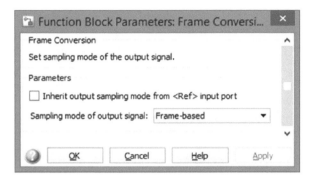

Figure 13.69 "Frame Conversion" Block Parameters.

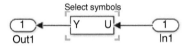

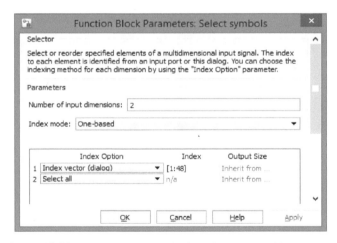

Figure 13.70 Remove Zeros Block and Associated Parameters.

The next block, labeled "Frame Conversion" produces a frame-based signal as shown in Figure 13.69.

The "Remove Zeros" block, shown in Figure 13.70 along with its parameters, produces a 48×20 array.

Figure 13.71 displays the "OFDM Symbols" block and its parameters used to convert the 48×20 array back to a 960-sample frame for insertion into the 16-QAM demodulator; the parameter settings are shown in Figure 13.72.

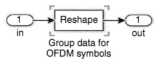

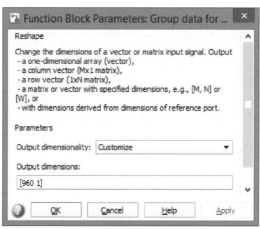

Figure 13.71 OFDM Symbols Block and Associated Parameters.

Figure 13.72 Rectangular 16-QAM Demodulator Parameters.

A summary of the principal OFDM Simulink model parameters used in Figure 13.58 are listed as follows:

- $M = 16, k = 4$
- Gray coding
- Sample time = 1 s
- Simulation time = 1,000,000 s
- Average signal power = 1 W
- Frame-based with 960 samples/frame
- 48×20 OFDM array
- 16 pad of zeroes
- 64-point IFFT and FFT
- 16-symbol cyclic prefix
- Receive and computation delay = 0
- Linear and nonlinear Saleh parameters

The Saleh parameters are now modified to cause the OFDM signal to remain in the nonlinear region of the power amplifier. Table 13.5 illustrates several arbitrarily selected Saleh parameters along with the corresponding BER assuming no additive noise.[4]

OFDM and QAM exhibit the same BER performance in AWGN, and for 16-QAM, the BER can be obtained from the QAM results in Chapter 4 as

$$P_b = \frac{3}{8}\text{erfc}\left(\sqrt{\frac{2\gamma_b}{5}}\right)\left\{1 - \frac{3}{8}\text{erfc}\left(\sqrt{\frac{2\gamma_b}{5}}\right)\right\}$$

As an example, with $E_b/N_o = \gamma_b = 10$ dB. $P_b = 0.0018$. When the OFDM transmission is operated within the linear region of the Saleh power amplifier with AM/AM = [1 0] and AM/PM = [0 0] using a 1,000,000 s simulation time, the error rate calculations shown in Figure 13.58 are OFDM BER = 0.001777 and the 16-QAM BER = 0.001735. It is then observed that both the OFDM BER and the 16-QAM BER are in close agreement with the theoretical BER.

Figure 13.73 displays the selected Saleh parameters as AM/AM = [1 0.1] and AM/PM = [0.05 0], which are used to determine the OFDM BER in the

---

[4]Note that the results provided in Chapter 4 were obtained with a QAM modulator whose input to the Saleh characteristic was normalized to unity. Thus the Saleh parameters were different from those chosen for this example.

TABLE 13.5 Saleh Parameters and Associated OFDM BER

Saleh	AM/AM	Saleh	AM/PM	BER
1	0	0	0	0.0
1	0	0.1	0	0.002
1	0	0.5	0	0.21
1	0.1	0.1	0	0.014
1	0.5	0.1	0	0.156
2	0.5	2	0.5	0.24

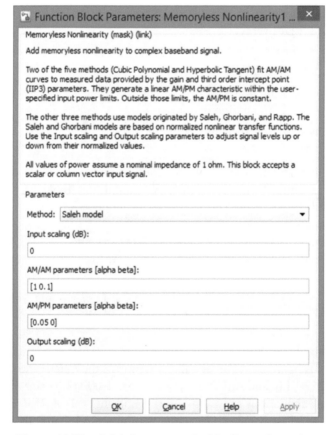

Figure 13.73 Saleh Parameters for Nonlinear Operation.

nonlinear case. The bertool is executed to obtain simulated OFDM BER performance for both linear and nonlinear operation with the results shown in Figure 13.74. This demonstrates that it is undesirable to operate an OFDM system in the nonlinear region of the power amplifier.

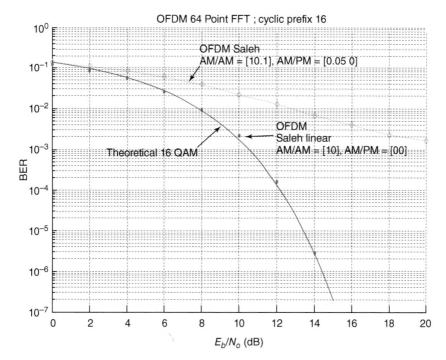

Figure 13.74 Theoretical 16-QAM BER and OFDM BER in Linear and Nonlinear Operation.

13.7 TURBO CODING WITH BPSK

Turbo codes have been adopted for use in many communications systems such as LTE, WIMAX, and satellite communications. The popularity of Turbo codes is due to their ability to approach the Shannon limit of $E_b/N_o = -1.6$ dB in AWGN, where reliable communications can be attained. As a practical case, rate 1/2 Turbo coding with BPSK modulation in an AWGN channel has been shown to yield a BER $=10^{-5}$ at $E_b/N_o = 0.7$ dB.[5] A brief description of Turbo coding is provided prior to introducing a Simulink model.

Turbo coding, also referred to as parallel concatenated convolutional codes (PCCC),[6] uses an encoder with two parallel, recursive systematic convolutional encoders, referred to as constituent encoders, interconnected by a random block interleaver followed by an output that ordinarily includes

---

[5] Sklar, B., "A Primer on Turbo Code Concepts", *IEEE Communication Magazine*, Dec. 1997, pp. 94–102.

[6] In general, PCCC may have more than two convolutional encoders that are not necessarily the same.

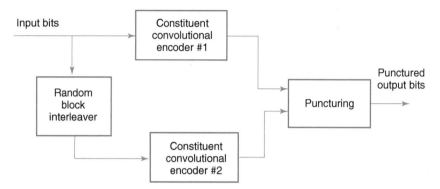

Figure 13.75 Turbo Encoder.

puncturing to remove systematic and parity bits, resulting in an increase in the overall code rate. The Turbo encoder just described is shown in Figure 13.75. This implementation produces a code having very few low-weight code words, and due to the interleaver, experiences relatively few nearest-neighbor errors.

The MathWorks documentation provides a specific example of a Turbo encoder that multiplexes the information stream with two identical encoder outputs where the Trellis structure for each encoder is described by poly2trellis(4, [13 15], 13). With a 64-bit input vector, the output consists of 64 systematic bits, 64 bits of parity from the first encoder, and 64 bits of parity from the second encoder, ending with 12 bits of tail for a total of 204 bits Thus the overall code rate is $64/204 \approx 1/3$.

In the Turbo decoder, soft input/soft output (SISO) decoding maximizes use of the received signal information where

- Multiple decoding operations are performed with soft information carried from one iteration to the next.
- An iterative computation of the *a posteriori* probability for each information bit is made with bit decisions becoming more certain at each iteration.
- When the iterations are stopped, the *a posteriori* probabilities are converted to hard decisions on information bits.

A generic example of a Turbo decoder is shown in Figure 13.76. The decoder functions include the following:

- Each decoder uses the *a posteriori* probabilities to form likelihood estimates of the information bits; then soft estimates are sent to the other decoder via extrinsic information extracted from parity bits.
- The decoder iterates until convergence is attained.
- A stopping rule is used to terminate iterations.

A Turbo code Simulink model adapted from MathWorks documentation is displayed in Figure 13.77 where BPSK modulation is modeled along with uncoded BPSK for comparison. Examining the masks in the Turbo encoder and decoder reveals that the encoder and decoder are system objects from MATLAB identified as "comm.TurboEncoder" and "comm.TurboDecoder." The constituent encoders are the same, supporting only rate-$1/N$ trellises where N is an integer. In this Simulink model, the trellis structure for each encoder is the same and is poly2trellis(4, [13 15 17], 13) where the constraint length is 4 and the remaining octal numbers identify the feedforward and feedback connections. Using a 200-sample frame length, it is observed that including tail bits, the overall code rate is $200/1018 \approx 1/5$. The Turbo decoder offers a choice of the true *a posteriori* probability decoding (True APP) or an approximate decoding algorithm for increased speed of computation. In generating the following results, only the True APP selection is used. The interleaver is implemented with randomly chosen indices using the MATLAB routine randperm. In the MATLAB command window, the interleaver indices are specified by entering ilv = randperm(200,200). For the results displayed in Figure 13.77, the E_b/N_o value is entered as $E_b/N_o = 2$, which results in BER = 0.001138 where the simulation is stopped after 100,000 s with over 100 errors incurred. Note that the uncoded BPSK BER = 0.03876, which is clearly much poorer than the Turbo code BPSK result.

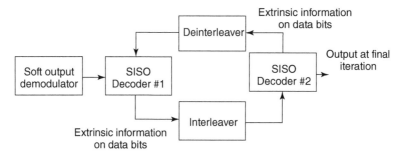

Figure 13.76 Generic Turbo Decoder.

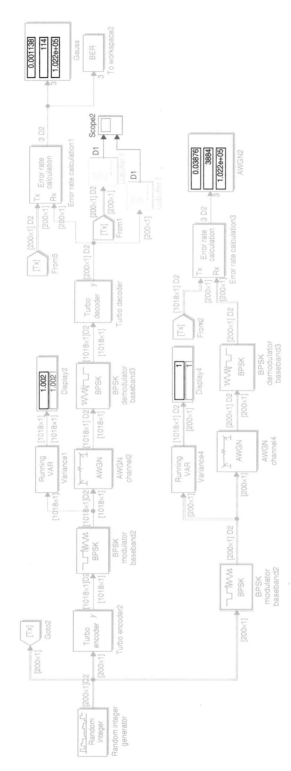

Figure 13.77 Turbo Coding Simulink Model Using BPSK Modulation ($E_b/N_o = 2$ dB).

TURBO CODING WITH BPSK

The Simulink model parameters used in Figure 13.77 are summarized as follows:

Model Parameters for Rate-1/5 Turbo Code with BPSK in AWGN

- BPSK antipodal signals $= +1$ and -1 ($M = 2$)
- Each constituent encoder: poly2trellis(4, [13 15 17], 13)
- Frame based with 200 samples/frame
- Sample time $= 1$ s
- Simulation time $= 100{,}000$ s
- Interleaver indices: ilv = randperm(200, 200)
- AWGN input signal power $= 1$ W with 200 s symbol period
- Turbo decoding algorithm: True APP
- Number of decoding iterations: 8
- Computation delay = receive delay = 0 s
- $E_b/N_o = 2$ dB

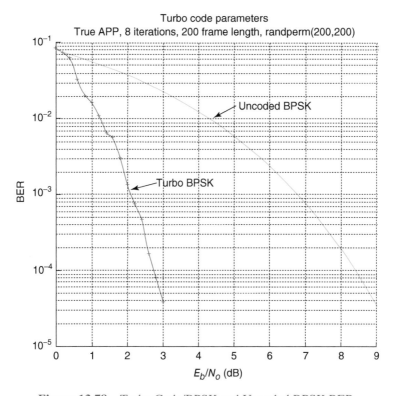

Figure 13.78 Turbo Code/BPSK and Uncoded BPSK BER.

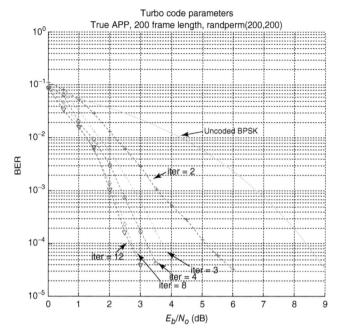

Figure 13.79 Turbo Code BPSK BER for Various Number of Iterations, Compared with BER for Uncoded BPSK.

TABLE 13.6 Frame Length Parameters

Frame Length	AWGN Symbol Period (s)	Interleaver Indices, ilv
200	200	randperm(200,200)
800	800	randperm(800,800)
2000	2000	randperm(2000,2000)

Figure 13.78 shows BER results for the Turbo code BPSK and uncoded BPSK where the improvement from Turbo coding relative to uncoded BPSK is apparent.

Figure 13.79 displays the simulated BER for Turbo BPSK where the number of iterations equals 2, 3, 4, 8, and 12 for a frame length set to 200. Uncoded BPSK BER is included for comparison.

Figure 13.80 shows the Turbo code BPSK BER performance obtained when using different interleaver specifications. The frame length is 200 samples and 8 iterations are used; the interleaver indices are specified as either ilv = (200:−1:1) or ilv = randperm(200, 200). Observe that in the case of the interleaver with indices ilv = (200:−1:1), pseudo-random characteristics are not exhibited, leading to degraded BER performance.

TURBO CODING WITH BPSK

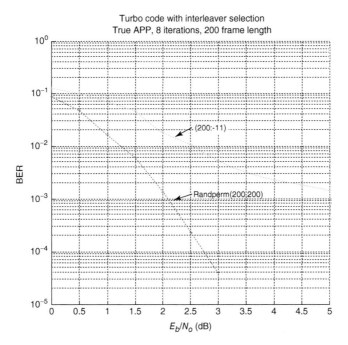

Figure 13.80 Turbo code BPSK BER Using Different Interleavers.

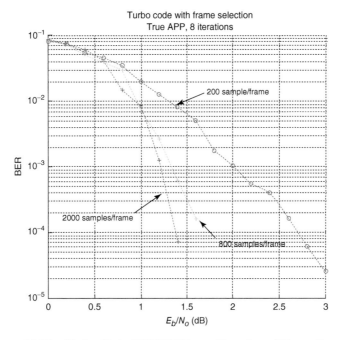

Figure 13.81 Turbo Code BPSK BER as a Function of Frame Length.

Figure 13.81 shows the Turbo code BPSK BER as a function of frame length. Table 13.6 provides the corresponding parameters for each frame length.

The models and performance results presented in this section have demonstrated that Turbo code BER performance with BPSK is a function of the selections of the constituent codes and the decoding algorithm. It has also been demonstrated that additional parameters including the number of iterations, the frame length, and the interleaver selection, have a significant impact on BER performance.[7]

---

[7]Further Simulink results on the Turbo code BER performance obtained in this section are available at: I. Raad and M. Yakan, "Implementation of a turbo code test bed in the Simulink environment," University of Wollongong Research Online, 2005.

APPENDIX A

PRINCIPAL SIMULINK BLOCKS USED IN CHAPTERS 1–13

A.A.1 SOURCES (FIGURE A.1)

Notes:

- The DSP sine wave block is convenient for vector-based computation where that computation method can be selected.
- The Bernoulli binary generator produces random ones and zeros.
- The Random integer generator produces M-ary random integers, typically used in conjunction with M-ary modulations.

A.A.2 SINKS (FIGURE A.2)

Notes:

- The To Workspace block sends data to the workspace for subsequent study.
- The display provides a view of the results following the Simulink execution.
- The scope block enables viewing multiple real-valued waveforms.

Modeling of Digital Communication Systems Using SIMULINK®, First Edition.
Arthur A. Giordano and Allen H. Levesque.
© 2015 John Wiley & Sons, Inc. Published 2015 by John Wiley & Sons, Inc.
Companion Website: www.wiley.com/go/simulink

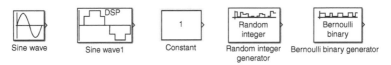

Figure A.1 Sources.

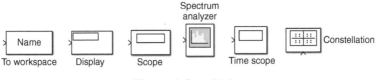

Figure A.2 Sinks.

- The Time scope from the DSP library accepts real or complex inputs.
- The spectrum analyzer has multiple settings including the FFT window size and type.

A.A.3 MODULATIONS (FIGURE A.3)

Notes:

- The Communications System Toolbox has an extensive set of digital baseband modulators and demodulators.
- Many Simulink models developed in this book use BPSK as a well-known modulation, where the focus is typically on another aspect of a complex simulation.

A.A.4 CHANNELS (FIGURE A.4)

Notes:

- The AWGN block in the Communications System Toolbox appears in most BER computations where the parameter E_s/N_o or E_b/N_o is the commonly used mode and the symbol period is available.
- The Gaussian noise generator allows the mean, variance and sample time to be specified.
- The Rayleigh and Rician fading channels allow multipath parameters to be specified.

SIGNAL PROCESSING (FIGURE A.6)   365

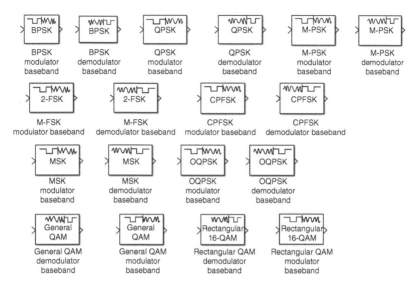

Figure A.3   Modulators and Demodulators.

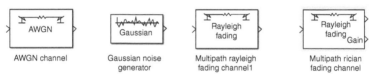

Figure A.4   Channels.

A.A.5   ERROR CONTROL CODING (FIGURE A.5)

Notes:

- The Error Correction and Detection section within the Communications System Toolbox allows the choice of several block error control codes and convolutional codes.
- In many system applications, matrix interleaving improves the effectiveness of error control coding.

A.A.6   SIGNAL PROCESSING (FIGURE A.6)

Notes:

- The DSP System Toolbox includes FFT and IFFT blocks.

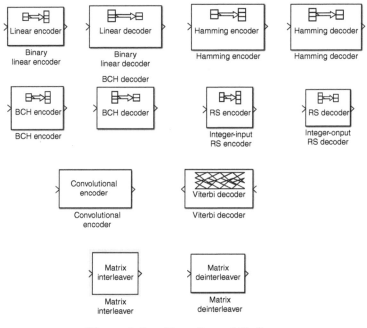

Figure A.5 Error Control Coding.

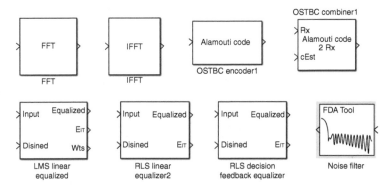

Figure A.6 Signal Processing.

- In the Communications System Toolbox, MIMO blocks include the OSTBC encoder and combiner.
- In the Communications System Toolbox, equalizer choices are available.

SPECIAL BLOCKS (FIGURE A.8)

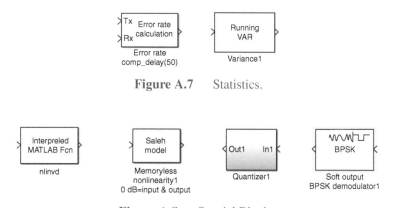

Figure A.7 Statistics.

Figure A.8 Special Blocks.

A.A.7 STATISTICS (FIGURE A.7)

Notes:

- In the Communications System Toolbox, the error rate block allows the computation and receive delay to be specified.
- The DSP System Toolbox contains a variance block that permits specialization to running variance for power estimation.

A.A.8 SPECIAL BLOCKS (FIGURE A.8)

Notes:

- User-defined embedded Simulink blocks are specified as interpreted MATLAB functions.
- In the Communications System Toolbox, under RF Impairments, multiple choices of the nonlinearity characteristic are available in the Memoryless Nonlinearity block.
- The Quantizer is found in the DSP System Toolbox.

APPENDIX B

FURTHER READING

Dabney, J. B., T. L. Harman, *Mastering Simulink*, Prentice Hall, Upper Saddle River, NJ, 2003.

Jamshidi, N., A. Farzad, B. Pedar, *Applied Guide to Simulink—Step by Step Tutorial*, LAP LAMBERT Academic Publishing, Saarbrüchen, Germany, 2010.

MathWorks Documentation (various) obtained by selecting the Help button.

Proakis J. G., M. Salehi, G. Bauch, *Contemporary Communication Systems using MATLAB and Simulink*, 2nd ed, Thomson Brooks/Cole, Pacific Grove, CA, 2004.

Silage, D., *Digital Communication Systems Using MATLAB® and Simulink®*, Bookstand Publishing, Gilroy, CA, 2009.

Modeling of Digital Communication Systems Using SIMULINK®, First Edition.
Arthur A. Giordano and Allen H. Levesque.
© 2015 John Wiley & Sons, Inc. Published 2015 by John Wiley & Sons, Inc.
Companion Website: www.wiley.com/go/simulink

INDEX

Adaptive Equalization, 247–283
  BPSK BER using LMS equalizer, 248, 254
  convergence, 251
  Decision Feedback Equalization (DFE), 268–273
  error signal, 249
  forward and feedback taps, 270
  Filter Design and Analysis Tool, 251
  fractional spacing, 248, 270
  intersymbol interference (ISI), 248
  Kalman equalizer, 344
  leakage factor, 258, 263
  LMS algorithm, 248, 251, 270
  LMS equalizer parameters, 262
  multipath and ISI, 248
  orthogonality principle, 249, 250
  QPSK scatter diagrams, 265
  Recursive Least Squares (RLS) equalizer, 273–280
  RLS algorithm definitions, 275
  RLS equalization in fading, 273, 275, 279
  stochastic gradient algorithm, 251
  tap coefficients, 248, 249
  training sequence, 251
Alamouti STBC:
  and BCH block coding, 210
  and QAM, 163, 219
  MIMO implementation, 165
  Simulink model, 157
Antenna nulling for interference cancellation, 313
Autocorrelation, 42
Automatic gain control (AGC), 122
AWGN block, 41, 44, 47, 51, 58, 364
AWGN parameter selections, 48, 50, 66, 174

Baseband modulation, 5, 364
BCH code:
  BER performance in AWGN, 171–175
  BER performance in fading, 194
  generator polynomial, 172
  interleaving, 195
  parameters, 172

Modeling of Digital Communication Systems Using SIMULINK®, First Edition.
Arthur A. Giordano and Allen H. Levesque.
© 2015 John Wiley & Sons, Inc. Published 2015 by John Wiley & Sons, Inc.
Companion Website: www.wiley.com/go/simulink

BCH code (*Continued*)
    punctured code, 172
    upper bound on code word error, 174
Binary PSK (BPSK), 5, 44
Bertool, 58
Bit-error rate (BER) computation, 44
Block information window, 8
Block libraries, 3
Buffering for BCH encoding, 172

Callbacks tab, 20
Changing a block label, 21
Channels:
    AWGN, 5, 41, 43–78
    Jakes, 120
    multipath, 127, 130
    Rayleigh, 119, 167
    Rician, 24, 130
Code rate, 167, 172, 179, 181, 186, 194, 201, 226, 237
Coding gain, 175, 177, 181, 185, 195, 204, 232
Coherent detection, 74, 79
Colors, 10, 12, 17, 22, 28
Command Window, 3, 16, 22, 81
Command History, 3
Communications System Toolbox, 1, 4
Connecting blocks, 8
Constant block, 4
Constellation display, 44
Continuous phase frequency shift keying (CPFSK), 108, 148
Convolutional coding and decoding, 225–244
    constraint length, 226
    feedback taps, 226
    free distance, 226
    generator polynomial, 226
    hard decision decoding, 226–229
    poly2trellis, 226–230, 356
    soft decision decoding, 229–233
    traceback depth, 227, 229
    transfer function, 229
    trellis structure, 227
    Viterbi decoder, 227
Copying a block, 6, 7, 8

Data buffering, 172
Data generators:
    Bernoulli binary generator, 5, 363
    random integer generator, 363
Data type, 12, 44, 46, 53
Decision Feedback Equalizer (DFE), 268
Delay:
    block, 4
    computation, 52
    decoding, 198
    element, 226, 240, 248
    receive, 52, 198
Digital baseband modulation, 5, 364
Display, 5
    scope parameters, 9, 10, 11
    tab, 12
Diversity transmission, 159, 201
Doppler Shift:
    diffuse, 136
    maximum, 120
Double precision numbers, 44, 46, 50
DSP implementations:
    ASIC or FPGA device, 70
    fixed point arithmetic, 70
    word size and fraction length, 70
DSP System Toolbox, 1, 4, 365
    complex signals, 44, 56
    quantizer, 26, 229, 367
    signal processing blocks, 365
    special blocks, 367
    variance block for power estimation, 367

Edit tab, 21
Energy Contrast Ratio (Eb/No), 56, 138
    symbol based Es/No, 56
Equalizer: see Adaptive Equalization
Error diagnostics, 24
Error rate calculation block, 44, 367
Error-control coding:
    BCH code, 171, 174
    convolutional code, 226
    Golay code, 172, 192, 195
    Hamming code, 175
    linear cyclic block code, 171, 173, 175
    punctured code, 172
    RS code, 181, 186, 207, 211
    simulation blocks available, 365
    table of BER results for selected codes, 191
    Turbo coding, 355
Error Rate Calculation block, 44, 52, 367
Eye (Identity) matrix, 180

INDEX **373**

Fading channel models:
 and interleaving, 195, 202, 204
 frequency-nonselective, 120
 Jakes model, 120, 125, 139, 194, 210
 multipath fading, 127, 130
 Rayleigh fading, 119, 167
 Rician fading, 124, 130
FFT and IFFT blocks, 365
Figure Properties, 21, 28
Filters:
 Finite Impulse Response (FIR), 248
 Kalman, 274, 275, 310, 320–342
Fixed-point arithmetic, 68
Flat Top window, 38
Floating point to fixed point conversion, 70
Frame-based Simulink model, 62
Frame-based vs. sample-based computation, 62
Free distance, 226
Frequency-Shift Keying (FSK):
 Binary FSK (BFSK), 101–107
 Continuous Phase FSK (CPFSK), 108
 M-ary FSK (MFSK), 107–108

Galois field, 176
Gaussian noise block, 59
Generator matrix for a linear block code, 171, 192, 195
Generator polynomial, 172, 218
Golay code:
 BER approximation, 192
 BER performance in AWGN, 179
 BER performance in fading, 195, 215
 bit error probability formulas, 199, 201
 generator matrix, 179, 195, 218
 parameters, 179
Gray coding, 64, 68, 81, 85, 87, 88, 343

Hamming code:
 BER performance in AWGN, 175–179
 generator matrix, 179
 parameters, 179
 primitive polynomial, 176
 upper bound on BER, 177
Hamming distance, 226
Hann window, 38, 108, 142
Hard-decision decoding, 174, 199, 226, 228

Identity (Eye) matrix, 180

Inherited precision, 46
Interference cancellation, 285, 291–298
 BER results with sinusoidal interference, 296
 RLS filter convergence, 273, 275
 RLS interference canceller, 291–298
 sinusoidal interference, 285, 291–296
Interleaving:
 and fading, 195
 interleaver length, 201
 in Turbo coding, 355
Intersymbol interference (ISI), 248, 343

Jakes fading model, 120
Jamming, 298

Kalman filtering, 6, 274, 285, 320, 322–328, 339
 aircraft tracking, 339–342
 equalizer convergence, 331, 332, 336, 338
 Extended Kalman Filter (EKF), 285, 339–342
 Kalman algorithm summary, 322
 Kalman and LMS equalizer comparison, 329, 331, 332, 336, 338
 Kalman equalizer, 285, 328–334, 336, 339
 Kalman gain, 321–324, 328, 332, 341
 Kalman vs. LMS convergence, 336, 338
 MATLAB Function for EKF, 341
 MATLAB Function for Kalman Equalizer Coefficients, 332
 MATLAB Function for LMS Equalizer Coefficients, 336
 radar tracking, 285, 339–342
 Scalar Kalman Filter, 285, 322–327

Labels, 10, 21
Least-Mean Square (LMS) equalizer, 248
Libraries in Simulink, 3
Linear cyclic block code, 175
Linear Predictive Coding (LPC), 286–291

M-ary Phase-Shift Keying (MPSK), 79
Masked block, 229, 343
MATLAB:
 and Simulink, 2
 Central File Exchange, 343

MATLAB (*Continued*)
  Command Window, 16
  Model Explorer, 19
  starting a MATLAB session, 2
MathWorks web link, 1
Minimum distance:
  BCH code, 172
  between QAM symbols, 85
  convolutional code, 226
  Golay code, 179
  Hamming code, 175
  Reed-Solomon code, 186
Minimum-Shift Keying (MSK), 108
  power spectrum, 113
  simulation model, 108
Model configuration parameters, 22
Model Explorer, 15, 17
Modulation and demodulation in AWGN:
  BFSK theoretical BER, 106
  BPSK and QPSK error rate performance, 43–78
  BPSK demodulator, 44
  BPSK fixed and floating point simulation, 70
  BPSK fixed point performance, 68
  BPSK theoretical BER, 56, 73, 159
  BPSK with phase offset, 75
  BPSK with STBC, 157–162
  CPFSK, 108
  error rate calculation block, 44
  FSK and MSK error rate performance, 101–113
  FSK modulator spectra, 101, 108
  FSK theoretical BER with noncoherent detection, 106, 108
  GMSK impulse response, 111
  MFSK simulation model, 107
  MPSK and QAM error rate performance, 79
  MPSK fixed point BER performance, 83
  MPSK theoretical BER performance, 82
  MSK and GMSK theoretical BER, 111
  OQPSK, 108
  QAM avergage power vs. peak power, 90
  QAM power amplifier constraint, 85
  QAM simulation model, 85
  QAM theoretical BER, 88, 163
  QPSK constellation, 67
  QPSK error rate performance, 64
  QPSK theoretical symbol error rate, 67
  Quadrature Amplitude Modulation (QAM), 79
  real and imaginary parts, 51
  signal constellation, 44, 64, 67, 83, 87
Modulation and demodulation in AWGN and Fading, 119–156
  Automatic Gain Control (AGC), 122
  BFSK BER performance in Rayleigh fading, 141–147, 151
  BFSK BER performance in Rician fading, 147
  BFSK in Rician fading and multipath, 148
  BFSK theoretical BER in Rayleigh fading, 144, 152
  BFSK theoretical BER in Rician fading, 147, 153
  BPSK BER performance in Rayleigh fading, 119–124
  BPSK in Rician fading and multipath, 127–137
  BPSK theoretical BER in Rayleigh fading, 120, 137
  BPSK theoretical BER in Rician fading, 124, 138
  CPFSK simulation model, 151
  CPFSK BER in Rician fading and multipath, 148–150
  Jakes model for flat fading, 120
  MFSK theoretical BER performance in Rayleigh fading, 142
  QAM in Rayleigh fading with STBC, 163, 167
  Rician channel Doppler spectrum, 127
  Rician channel impulse response, 126
  Rician K factor, 126
Modulation with block coding, 171–191, 193–223
  BPSK and BCH coding in AWGN, 171–174, 177, 191
  BPSK and BCH coding in fading with interleaving, 193–199
  BPSK and BCH coding with Alamouti STBC, 210, 213–216
  BPSK and Golay code in AGWN, 171, 179–183, 191

INDEX 375

BFSK and Golay code in fading, 193, 195, 198–202
BFSK and Golay code with Alamouti STBC, 215–219
BFSK and Golay code with diversity, 204
BPSK and Hamming code in AWGN, 171, 175–180, 191
FSK and RS coding in AWGN, 171, 181, 183–185, 191
FSK and RS coding in fading, 193, 201–206
FSK and RS coding with Alamouti STBC, 193, 218–221
QAM and RS coding in AWGN, 171, 186–188, 191
QAM and RS coding in fading, 193, 204, 206–212
QAM and RS coding in multipath, 188–190
QAM and RS coding with Alamouti STBC, 194, 219, 221–223
Modulation with convolutional coding, 225–244
    BPSK and hard decision decoding in AWGN, 225–230
    BPSK and soft decision decoding in AWGN, 225, 229–233
    BPSK and convolutional coding in AWGN and fading, 233–239
    BPSK with Alamouti STBC in fading, 225, 239–243
    changing the traceback depth, 227–230
    hard decision Viterbi decoding, 227–230
    quantizer block, 229–242
    soft decision Viterbi decoding, 229–233
    theoretical performance for binary symmetric channel, 228–229
    upper bound on bit error probability, 229
Multipath channel, 5, 127, 180, 188
Multipath interference in Rician fading, 131
Multiple-Input Multiple-Output (MIMO):
    MIMO and OSTBC blocks, 366
    with BPSK and STBC, 165
    with QAM in fading, 163

Noise :
    AWGN channel block, 5, 41, 44, 48, 51
    complex noise variance, 59
    Gaussian Noise block, 59

    generators, 5
Noncentrality parameter in Rician fading model, 153
Noncoherent BFSK, 101, 141, 147, 152
Noncoherent MFSK, 107, 142, 153
Nonlinear power amplifier, 91
Normal equations, 250
Number of bit errors, 52
Nyquist sampling, 32

Octal representation of feedback taps, 226
Offset QPSK (OQPSK), 108
Orthogonal Frequency Division Multiplexing (OFDM), 343
    BER for OFDM and QAM, 343, 344, 353–355
    cyclic prefix, 343–345, 348–350, 353, 355
    Fast Fourier Transform (FFT), 343, 344, 349, 350, 353, 355
    Inverse FFT (IFFT), 344, 345, 347, 348, 353
    LTE, 343
    rectangular 16-QAM modulator, 343, 344, 346, 352
    Saleh model parameters, 344, 345, 349, 353–355
    WiMAX, 343
Output block, 19, 44

Periodogram method, 37
poly2trellis, 226–230, 356
Power amplifier constraint:
    AM/PM conversion, 92
    predistortion block, 96
    Saleh nonlinear model, 92, 96, 345
Primitive polynomial, 172, 176
Principal Simulink Blocks, 363–367

Quadrature Amplitude Modulation (QAM), 5, 85–98
Quantizer, 229, 367

Random number generators:
    Bernoulli binary generator, 5, 363
    Random integer generator, 363
    Random integer seed, 44
    Random integer source block, 44
Rayleigh fading model, 5, 119

Recursive Least Squares (RLS) equalizer, 273
Reed-Solomon (RS) code:
  parameters, 178
  BER performance with FSK in AWGN, 181, 185
  BER performance with QAM in AWGN, 181, 186
Renaming a model, 6
Resolution bandwidth, 103
Rician K factor, 124
Routing symbols, 47
Run time, setting, 28, 42
Running a simulation, 11, 15
Running Variance block, 28, 40, 59, 367

Saleh model for power amplifier, 92
Sample based computation, 16
Sample based signals, 16
Sample time, 12
  and sample rate, 56
  legend, 12
Saving a model, 19
Scatter plots, 64, 68, 93, 262
Scopes:
  display, 10
  setting parameters, 9
  Scope parameters window, 10
  history page, 10
  time scope, 103, 364
Sending data to Workspace, 18
Signal power, 28, 40, 48, 53, 60
Signal Processing blocks, 365
Signal-to-noise ratio, 56, 138
Simulation menu, 9, 20
Simulink:
  block libraries, 3, 363–368
  building a new Simulink model, 6
  commonly used blocks, 4
  Communications System Toolbox, 5
  DSP System Toolbox, 5
  examples, 285–362
  executing the Simulink model, 11
  inserting Signal Source and Scope, 6
  model-wide utilities, 8
  model info, 8
  setting Scope parameters, 9
  setting Source block parameters, 8
Simulink Library Browser, 7
  using Model Explorer, 19
Sine Wave:
  block, 8, 12, 15, 19, 24
  phase shift, 32
  Simulink model, 27
  Source block parameters, 28
  spectrum, 27, 32
Sink block parameter, 30
Solvers:
  solver choices, 23
  shortened simulation time, 23
Sources, 5, 6, 7, 41, 363
Space-Time Block Coding (STBC), 157–168, 210–223
  BCH BPSK, 210, 213–216
  BPSK and convolutional coding in fading, 239–243
  BPSK BER, 167
  BPSK in fading, 157–161
  FSK and RS coding, 218–221
  Golay BFSK, 215–219
  QAM and RS coding, 219, 221–223
  QAM BER, 167–168
  QAM in fading, 163
  STBC algorithm, 165–167
Spectrum analyzer:
  FFT, 38, 103
  Flat Top window, 38
  Hann window, 34, 113
Spectrum Scope, 103, 108
Speech Compression, 286–291
  all-pole vocal-tract model, 286, 289
  Levinson-Durbin algorithm, 287
  Linear Predictive Coding, 286
Spread Spectrum, 288–313
  excision, 309–313
  in-band interference, 303, 309
  theoretical BER, 303
Stop time, 23
Symbol error rate, 67, 81, 88, 108, 185, 265
Synchronizing a simulation, 234, 249

Text box, 8
Timeseries format, 28
Toolboxes, 1, 4, 5, 103, 157, 265, 364
Traceback depth for MSK and GMSK, 113
Transfer function for a convolutional code, 229
Turbo coding, 355–362

BER for BPSK, uncoded and
    Turbo-coded, 359
  decoding iterations, 356, 360
  frame length, 357
  SISO decoding, 356

Utility block, 103

Variables, defining, 3, 16
Variance block, 6, 28, 40
Vector-based computation, 363

Vector inputs, 14
Viterbi decoding:
  hard-decision, 226–229
  soft-decision, 229–233

Window:
  Command, 3, 16, 21, 81, 83, 349
  Model, 6, 8, 11, 22
Workspace block, 28
Workspace variables, 30

CPSIA information can be obtained
at www.ICGtesting.com
Printed in the USA
BVHW020222041019
560060BV00007B/1/P